AF678962

Dyslexia and Neuroscience

The DYSLEXIA Foundation

The Extraordinary BRAIN series

Dyslexia and Neuroscience
The Geschwind-Galaburda Hypothesis 30 Years Later

edited by

Albert M. Galaburda, M.D.
Harvard Medical School/Beth Israel Deaconess Medical Center
Boston, Massachusetts

Nadine Gaab, Ph.D.
Boston Children's Hospital, Harvard Medical School
Boston, Massachusetts

Fumiko Hoeft, M.D., Ph.D.
University of California, San Francisco

and

Peggy McCardle, Ph.D., M.P.H.
PM Consulting, LLC, and Haskins Laboratories
Tarpon Springs, Florida

with invited contributors

PAUL·H· BROOKES PUBLISHING CO®

Baltimore • London • Sydney

·P A U L·H·
BROOKES
PUBLISHING CO.®

Paul H. Brookes Publishing Co.
Post Office Box 10624
Baltimore, Maryland 21285-0624
USA
www.brookespublishing.com

Copyright © 2018 by Paul H. Brookes Publishing Co., Inc.
All rights reserved.

"Paul H. Brookes Publishing Co." is a registered trademark of
Paul H. Brookes Publishing Co., Inc.

Typeset by BMWW, Windsor Mill, Maryland.
Manufactured in the United States of America by Sheridan Books, Inc., Chelsea, Michigan.

The views expressed in this book are those of the authors and do not necessarily represent those of the National Institutes of Health, March of Dimes, or the Government of the Russian Federation.

Library of Congress Cataloging-in-Publication Data

Names: Galaburda, Albert M., 1948– editor.
Title: Dyslexia and neuroscience: the Geschwind-Galaburda hypothesis 30 years later / edited by Albert M. Galaburda, M.D., Harvard Medical School/Beth Israel Deaconess Medical Center, Boston, Massachusetts, Nadine Gaab, Ph.D., Boston Children's Hospital, Harvard Medical School, Boston, Massachusetts, Fumiko Hoeft, M.D., Ph.D., University of California, San Francisco and Peggy McCardle, Ph.D., M.P.H., PM Consulting, LLC, and Haskins Laboratories, Tarpon Springs, Florida; with invited contributors.
Description: Baltimore: Paul H. Brookes Publishing Co., [2017] | Series: The extraordinary brain series ; 15 | Includes bibliographical references and index.
Identifiers: LCCN 2017021915| ISBN 9781681252254 (hardback) | ISBN 9781681252605 (epub) | ISBN 9781681252636 (pdf)
Subjects: LCSH: Dyslexia--Research. | Brain--Diseases--Research. | BISAC: MEDICAL / Neuroscience. | PSYCHOLOGY / Developmental / General. | SCIENCE / Life Sciences / Neuroscience. | EDUCATION / Special Education / Learning Disabilities.
Classification: LCC RC394.W6 D9535 2018 | DDC 362.19685/5300724--dc23 LC record available at https://lccn.loc.gov/2017021915

British Library Cataloguing in Publication data are available from the British Library.

2021 2020 2019 2018 2017

10 9 8 7 6 5 4 3 2 1

Contents

About the Figures

Purchasers of this book may view online full color versions of noted grayscale figures at www.brookespublishing.com/galaburda/materials for both print and e-book buyers.

About the Editors

Albert M. Galaburda, M.D., Emily Fisher Landau Professor of Neurology and Neuroscience, Harvard Medical School/Beth Israel Deaconess Hospital Medical Center, Boston, Massachusetts

Dr. Galaburda is a cognitive neurologist at Beth Israel Deaconess Medical Center (BIDMC) and, since 1995, the Emily Fisher Landau Professor of Neurology (Neuroscience) at Harvard Medical School. In addition, he serves as Director, Office for Diversity, Inclusion, and Career Advancement at BIDMC and as Codirector, Mind, Brain, and Behavior Interfaculty Initiative at Harvard University. Dr. Galaburda is a native of Chile and a graduate of Boston University's Six-Year Liberal Arts-Medicine Program (1971). After medical school, he trained in internal medicine and neurology at Boston City Hospital (now Boston Medical Center). He directed a National Institutes of Health-funded bench laboratory on the fundamental causes of learning disorders, especially language-based learning disabilities, from 1979 to 2015, and he directed the Division of Cognitive Neurology at BIDMC from 1992 to 2015. He has written extensively on cerebral lateralization and dyslexia. He has published more than 230 original articles, reviews, book chapters, and books and has lectured extensively locally, nationally, and abroad on the general field of cognitive neurology. He has also been recognized for his work with several prizes, including the Pattison Prize in Neuroscience, the IPSEN Prize in Neural Plasticity, the American Academy of Neurology Decade of the Brain keynote speaker award, the Behavioral Neurology Society of the American Academy of Neurology Lifetime Achievement Award, the International Dyslexia Association's Samuel T. Orton Award, and others. Since January 2015, Dr. Galaburda devotes half of his time to diversity and inclusion as well as the career advancement of students and physicians underrepresented in medicine.

Nadine Gaab, Ph.D., Associate Professor of Pediatrics at Boston Children's Hospital, Harvard Medical School, Division of Developmental Medicine, Department of Medicine, Laboratories of Cognitive Neuroscience, and Faculty, Harvard Graduate School of Education, Boston, Massachusetts

Dr. Gaab received a doctoral degree in psychology from the University of Zurich in Switzerland and did postdoctoral training at Stanford University and the Massachusetts Institute of Technology. She joined the faculty at Boston Children's Hospital and Harvard Medical School in 2007. Her current research within the Laboratories of Cognitive Neuroscience addresses contemporary challenges of clinical practice and education with neuroscientific methods from infancy to adulthood. Her work primarily focuses on auditory, language, and music processing in the human brain and its applications for the development of typical and atypical language and literacy skills. The Gaab Lab utilizes structural and functional magnetic resonance imaging as well as behavioral measurement tools and is currently working on various research questions such as 1) Which brain learns to read best under which circumstances? 2) How do environmental factors influence the brain's ability to read? 3) Can neuroscience help improve the early identification of children at risk for dyslexia, and 4) What factors are important for shaping a resilient (reading) brain? The Gaab Lab employs cross-sectional and longitudinal study designs and works closely with more than 20 private and public schools within the greater New England area. Please visit http://www.gaablab.com for more information.

Fumiko Hoeft, M.D., Ph.D., Associate Professor in the Department of Child and Adolescent Psychiatry and Weill Institute for Neurosciences, University of California, San Francisco (UCSF), California

Dr. Hoeft is the Director of the Multi-UC-Campus Science-Based Innovation in Learning Center and UCSF Laboratory for Educational Neuroscience (http://www.brainLENS.org) and a research scientist at Haskins Laboratories. She is a member of the UCSF Dyslexia Center Board, the International Dyslexia Association (IDA) Board, the National Center for Learning Disabilities scientific advisory board, and the Center for Childhood Creativity scientific advisory board. She was the 2014 Geschwind Memorial Lecturer for for the IDA. BrainLENS focuses on how cognitive science can inform educational and clinical practices; more specifically, in understanding the neurobiological causes of dyslexia, early identification and prediction, and the emotional resilience necessary to succeed.

Peggy McCardle, Ph.D., M.P.H., Consultant, Peggy McCardle Consulting and Haskins Laboratories, New Haven, Connecticut

Dr. McCardle is a private consultant, science writer and editor, and affiliated research scientist at Haskins Laboratories. She is former chief of the Child Development and Behavior Branch of the *Eunice Kennedy Shriver* National Institute of Child Health and Human Development (NICHD), U.S. National Institutes of Health, where she also directed the research program in language, bilingualism, and biliteracy, and developed various literacy initiatives. She is a linguist, former speech-language pathologist, and classroom teacher and is the recipient of various awards for her work in federal government, including an NICHD Mentor Award. She received the Einstein Award from The Dyslexia Foundation in 2013. Her publications address aspects of public health, developmental psycholinguistics (e.g., language development, bilingualism, reading, learning disabilities), and education (especially reading and educational services for English language learners). She has extensive experience developing and co-editing volumes and thematic journal issues.

About the Contributors

Irene Altarelli, Ph.D., University of Geneva, Switzerland

Dr. Altarelli completed her doctorate in Paris under the supervision of Franck Ramus and Ghislaine Dehaene. She used magnetic resonance imaging in her doctoral work to investigate the neuroanatomical correlates of developmental dyslexia, revealing sex-related differences in the dyslexic population. She has been a postdoctoral fellow in the Brain and Learning Laboratory, Faculté de Psychologie et des Sciences de l'Éducation at the University of Geneva since 2014, thanks to a Marie Curie postdoctoral grant, focusing on interindividual variability in learning abilities from both a cognitive and a neural perspective, under the supervision of Daphne Bavelier.

Emily Barrett, Ph.D., Rutgers University, New Brunswick, New Jersey

Dr. Barrett is an Associate Professor in the Environmental and Occupational Health Sciences Institute at Rutgers University. Her work focuses on the early origins of health and disease, or how exposures early in life shape our subsequent health and developmental trajectories. Much of her research focuses on prenatal exposure to endocrine disruptors, agents that interfere with the normal activity of hormones in the body.

Courtney Bodge, Ph.D., Butler Hospital, Providence, Rhode Island

Dr. Bodge is the Research Coordinator at Butler Hospital. In her doctoral work on sex differences in rodent models of neonatal brain injury, under Holly Fitch, University of Connecticut, she demonstrated differences in both degree of injury and long-term behavioral outcome, suggesting that potential therapeutic targets may vary for males and females. Subsequently at the Women & Infants Hospital and the Alpert Medical School at Brown University, her work focused on the permeability of the blood-brain barrier and neuroprotective strategies. She currently leads multiple clinical research trials at Butler Hospital, including studies of early onset Alzheimer's disease targeting specific gene mutations.

Laurie Cutting, Ph.D., Vanderbilt University, Nashville, Tennessee

Dr. Cutting is the Patricia and Rodes Hart Endowed Professor, Peabody College of Education and Human Development, and holds faculty appointments in the Departments of Special Education, Psychology, Radiology, and Pediatrics at Vanderbilt. She is also a senior scientist at Haskins Laboratories (New Haven, CT) and holds an adjunct faculty position at Johns Hopkins School of Medicine, Department of Neurology (Baltimore, MD). Her research focuses on educational neuroscience, particularly the neurobiological and behavioral underpinnings of reading, oral language, and dyslexia.

Lou Scotto di Covella, Ph.D. student, Centre National de la Recherche Scientifique (CNRS), Paris, France

Ms. Scotto di Covella is a doctoral student at the CNRS, in the Department d'Études Cognitives—École Normale Supérieure, Laboratoire de Sciences Cognitives et Psycholinguistique, Pierre and Marie Curie University (Paris), where she is working with Franck Ramus on the neuroanatomical correlates of developmental dyslexia. More precisely, she studies the brain foldings and cortical sulci using magnetic resonance imaging data from different countries.

Geert J. de Vries, Ph.D., Georgia State University, Atlanta

Dr. de Vries is Professor and Director of the Neuroscience Institute at Georgia State University. He is a past president of the Organization for the Study of Sex Differences as well as the Society for Behavioral Neuroendocrinology. Ever since discovering the sexually dimorphic nature of vasopressin innervation of the brain as a graduate student, De Vries has studied the development and function of sex differences in the brain. This culminated in proposing the overarching idea that such differences cause as well as prevent sex differences in physiology and behavior. He currently studies the effects of inflammation and microbiota on the brain.

Heidi M. Feldman, M.D., Ph.D., Stanford University School of Medicine, California

Dr. Feldman is the Ballinger-Swindells Endowed Professor of Developmental-Behavioral Pediatrics at Stanford University School of Medicine and the Director of Developmental-Behavioral Pediatrics Clinical Programs at Stanford Children's Health. As a researcher, she has focused on language and reading outcomes of children born preterm since the late 2000s. Her studies integrate traditional and novel behavioral assessments with advanced neuroimaging techniques. As a clinician, she summarized her approach in a book titled, *Redesigning Health Care for Children with Disabilities.* She argues that strengthening inclusion and functional outcomes should be the priority for health care for all children with disabilities.

R. Holly Fitch, Ph.D., University of Connecticut, Storrs

Dr. Fitch is Professor of Behavioral Neuroscience in the Department of Psychological Sciences, the Institutes for Systems Genomics and Brain and Cognitive Sciences, and Director of the Murine Behavioral Phenotyping Facility. Her research focuses on rodent models of neurodevelopmental disruptions and cognitive disabilities, including genetic risk factors for cognitive disability, animal models of brain damage typical of premature/term birth insult, and behavior studies of effects of prenatal teratogens. A variety of behavioral and anatomic assessments are employed, with an emphasis on tasks that tap skills foundational to communicative processes in rodents.

Elena L. Grigorenko, Ph.D., University of Houston and Baylor College of Medicine, Texas

Dr. Grigorenko is the Hugh Roy and Lillie Cranz Cullen Distinguished Professor, Department of Psychology, at the University of Houston and Professor, Departments of Pediatrics and Molecular and Human Genetics, at Baylor College of Medicine. Dr. Grigorenko and her laboratory have contributed to numerous studies of neurodevelopmental disorders, including language and reading disorders, conduct disorder, and autism spectrum disorder. Dr. Grigorenko and her colleagues are particularly interested in the etiology and developmental trajectories of these disorders.

Roeland Hancock, Ph.D., University of California, San Francisco

Dr. Hancock is a postdoctoral researcher in the Department of Psychiatry at the University of California, San Francisco, with broad interests in the neurobiology of language. His current research focuses on human auditory processing, the role of neurochemistry in regulating neural oscillations during speech processing, and how these processes may be disrupted in atypical language and literacy development. His other interests include imaging genetics and the use of mobile technology for the early identification of children at risk for dyslexia.

David S. Hong, M.D., Stanford University School of Medicine, California

Dr. Hong is Assistant Professor of Psychiatry at the Stanford University School of Medicine and Associate Director of Clinical Neuroscience in the Division of Interdisciplinary Brain Sciences. His current research examines the role of sex-specific factors on cognitive development, utilizing multimodal neuroimaging and genomic methods. His recent work has focused on neurodevelopmental trajectories in youth with sex chromosome aneuploidies and disorders of sexual development. Hong also directs clinical activities within the Division of Interdisciplinary Brain Sciences to advance the treatment and diagnosis of children with a broad range of neurodevelopmental disorders.

Katarzyna Jednoróg, Ph.D., Polish Academy of Sciences, Warsaw, Poland

Dr. Jednoróg is Adjunct Professor, Laboratory of Psychophysiology, Nencki Institute of Experimental Biology, at the Polish Academy of Sciences, where she teaches and conducts research. Her current research focuses on structural and functional neural correlates of developmental dyslexia and its early predictors across different orthographies. She is also involved in a longitudinal project on modality general and modality specific neuronal effects of second language learning.

Sergey A. Kornilov, Ph.D., University of Houston, Texas

Dr. Kornilov is the Duncan Scholar in Molecular and Human Genetics, Baylor College of Medicine, and Postdoctoral Fellow at the University of Houston. His work focuses on the neurobiological underpinnings of cognitive development, working with various populations (typically developing children and undergraduates, special and clinical populations) using a variety of approaches (from psychometric to molecular genetic). He received the Society for Research on Child Development Outstanding Doctoral Dissertation in Developmental Science Award in 2014 and the Isabelle Liberman Award in 2013 for his work on molecular and neurophysiological bases of developmental language disorders.

Nicole Landi, Ph.D., University of Connecticut, Storrs, and Haskins Laboratories, New Haven, Connecticut

Dr. Landi is an Assistant Professor of Psychological Sciences at the University of Connecticut and the Director of EEG Research at Haskins Laboratories. Dr. Landi's research seeks to better understand typical and atypical reading and language development through the use of multiple cognitive neuroscience methodologies (MRI, EEG) and neurogenetic analyses. Through this work her lab hopes to identify neurobiological and environmental mechanisms that contribute to individual differences in reading and language skill and to the complex etiology of disorders such as dyslexia, specific comprehension deficit and specific language impairment.

Eileen Luders, Ph.D., Cousins Center for Psychoneuroimmunology, Semel Institute for Neuroscience and Human Behavior, Department of Psychiatry and Biobehavioral Sciences, University of California, Los Angeles, School of Medicine

Dr. Luders is an Associate Professor of Psychiatry at the University of California, Los Angeles (UCLA). Trained in neuroscience and neuropsychology, she received her Ph.D. from the University of Zurich, Switzerland. Her research is focused on understanding the human brain, in health and disease, using high-resolution magnetic resonance imaging and state-of-the-art brain mapping techniques. Central themes of her studies are the brain's sexual dimorphism, brain development and aging, as well as structure-function relationships with emphasis on handedness, intelligence, and mindfulness.

Margaret M. McCarthy, Ph.D., University of Maryland School of Medicine, Baltimore

Dr. McCarthy is Professor and Chair, Department of Pharmacology, Program in Neuroscience, at the University of Maryland School of Medicine. She received her doctorate from the Institute of Animal Behavior at Rutgers University and received postdoctoral training at Rockefeller University and the National Institutes of Health as a National Research Council Fellow. McCarthy has been with the faculty of the University of Maryland School of Medicine since 1993 and became the Chair of the Department of Pharmacology in 2011. She has also served as Director of Graduate Education for the Program in Neuroscience and as Associate Dean for graduate education.

Tuong-Vi Nguyen, M.D., M.Sc., McGill University and McGill University Health Center, Montréal, Canada

Dr. Nguyen completed a master's degree and residency in psychiatry at McGill University as well as a postdoctoral fellowship at the National Institutes of Health in the Section on Behavioral Endocrinology and the Section on Integrative Neuroimaging. She currently holds faculty positions at McGill University, jointly appointed in the departments of psychiatry and obstetrics-gynecology, and is the Director of the Reproductive Psychiatry Program at the McGill University Health Center. Her research focuses on investigating the central nervous system effects of steroid hormones, particularly androgens, during development as well as during reproductive stages.

Sebastian Ocklenburg, M.Sc., Ph.D., Ruhr-University, Bochum, Germany

Dr. Ocklenburg is a lecturer in the Institute for Cognitive Neuroscience, Department of Biopsychology, with the Faculty of Psychology at Ruhr-University, Bochum, Germany. His research is in biopsychology, with a focus on the ontogenesis of hemispheric asymmetries. He uses molecular genetics and neuroimaging techniques to uncover how handedness and language lateralization originate.

Tom O'Connor, Ph.D., University of Rochester Medical Center, New York

Dr. O'Connor is Professor in the Department of Psychiatry and Director of the Wynne Center for Family Research at the University of Rochester Medical Center. His clinical research focuses on the role that early (including prenatal) exposures and experiences play in shaping psychological, physiological, and immunological processes underlying behavioral and somatic health.

Ola Ozernov-Palchik, M.S., Tufts University, Medford, Massachusetts

Ms. Ozernov-Palchik is a doctoral candidate at the Center for Reading and Language Research in the Tufts University's interdisciplinary cognitive science program under the mentorship of Maryanne Wolf. In her research, she is interested in understanding the cognitive and neural mechanisms of deficit and compensation in developmental dyslexia. She also conducts research with Nadine Gaab at Boston Children's Hospital and John Gabrieli at the Massachusetts Institute of Technology using neuroimaging, psychoeducational, and cognitive methods as part of several longitudinal studies of children at risk for dyslexia. Ozernov-Palchik is investigating the association between musical rhythm processing and early literacy development with Aniruddh Patel at Tufts University.

Silvia Paracchini, Ph.D., University of St Andrews, Scotland

Dr. Paracchini holds a Royal Society University Research Fellowship at the University of St. Andrews where she leads the Neurogenetics Group and the Bioinformatics Unit in the School of Medicine. She is a human geneticist whose research is aimed at understanding the biological component of dyslexia and combines genetic mapping (using genome-wide association studies and next generation sequencing) with functional characterization of candidate genes in biological models that include neuronal stem cells and zebrafish. More recently she has become interested in the genetics of handedness and how shared biology might influence both handedness and dyslexia. Dr. Paracchini is a member of the Royal Society of Edinburgh Young Academy of Scotland.

Jutta Peterburs, Ph.D., University of Münster, Germany

Dr. Peterburs studied psychology at Ruhr-University in Bochum, Germany, and received her doctorate in psychology in early 2012. She currently is a Postdoctoral Fellow at the Institute of Medical Psychology and Systems Neuroscience at the University of Münster, and she divides her time between Germany and the United States, where she is a Research Associate in the Cognitive Neuroscience Division at the Johns Hopkins School of Medicine. Her research focuses on modulators and neural correlates of performance monitoring in humans and nonmotor functions of the cerebellum.

Franck Ramus, Ph.D., Centre National de la Recherche Scientifique (CNRS), Paris, France

Dr. Ramus is Research Director and senior research scientist at CNRS and serves as Adjunct Professor at the École Normale Supérieure. He works at the Laboratoire de Sciences Cognitives et Psycholinguistique, Department of Cognitive Studies, École Normale Supérieure in Paris, where he heads the cognitive development and pathology team. His research bears on the development of language and social cognition in children, related disorders (developmental dyslexia, specific language impairment, and autism spectrum disorder), their cognitive and neural bases, and their genetic and environmental determinants.

John L.R. Rubenstein, M.D., Ph.D., University of California, San Francisco

Dr. Rubenstein is the Nina Ireland Distinguished Professor of Child Psychiatry at the Nina Ireland Laboratory of Developmental Neurobiology and Professor of Psychiatry at the University of California San Francisco Weill Institute for Neurosciences. He was recently awarded the Ruane Prize for Outstanding Achievement in Child and Adolescent Research from the Brain and Behavior Research Foundation. His research focuses on the regulatory genes that orchestrate development of the cerebral cortex and basal ganglia. These genes serve as entry points to elucidate human neuropsychiatric disorders.

Ana Vallejo Sefair, University of Rochester, New York

Ms. Vallejo Sefair is a graduate student in clinical psychology at the University of Rochester. Her work focuses on the early origins of behavioral and somatic health in children and adolescents.

Amanda Smith, Ph.D., Alkermes, Inc., Watertown, MA

Amanda Smith is a medical writer at Alkermes, Inc., a pharmaceutical company focused on developing treatments for central nervous system diseases. Prior to beginning a career in pharmaceuticals, she conducted her doctoral research at the University of Connecticut within the Department of Psychological Sciences, Behavioral Neuroscience Division. Working under the supervision of Dr. Holly Fitch, she designed a compilation of studies to assess the behavioral consequences following induced hypoxic ischemic brain injury in a rodent model. Her research was specifically focused on sex differences following injury and possible strategies to prevent or ameliorate associated long-term behavioral deficits.

Caitlin Szalkowski, Ph.D., Hilbert College, Hamburg, New York

Dr. Szalkowski is Assistant Professor of Psychology at Hilbert College. She studied neurobehavioral outcomes following genetic disruptions in a rodent model while working under Holly Fitch at the University of Connecticut. Her work demonstrated different patterns in behavioral and anatomical disruption following reduced expression of rodent homologs of candidate dyslexia susceptibility genes *Dyx1c1* and *Kiaa0319,* as well as evidence of sex disparity in reaction to these insults. This suggests possible mechanisms for the observed differences in rates of dyslexia among males and females as well as the varied presentation of specific deficits within the disorder.

David K. Urion, M.D., F.A.A.N., Boston Children's Hospital, Massachusetts

Dr. Urion received his undergraduate degree from Dartmouth College and his medical degree from Stanford University. He trained in internal medicine at the Peter Bent Brigham Hospital, in pediatrics at Boston Children's Hospital, and in child neurology in the Longwood Area Neurology Training Program. He serves as the director of education and residency training programs in child neurology and neurodevelopmental disabilities, as well as the director of behavioral neurology clinics and programs, at Boston Children's Hospital, where he holds the Charles F. Barlow Chair in Neurology. He is the immediate past president of the Professors of Child Neurology.

Xi Yu, Ph.D., Boston Children's Hospital and Harvard Medical School, Massachusetts

Dr. Yu is a postdoctoral fellow in the Laboratories of Cognitive Neuroscience at Boston Children's Hospital. Her research interests lie in the cognitive and neural mechanisms underlying language and reading development in typical and atypical populations. Her work also seeks to identify biomarkers for early diagnosis of dyslexia and compensatory mechanisms in children, which could inform the design of pre-reading treatment for children at risk for dyslexia.

Jingjing Zhao, Ph.D., Shaanxi Normal University and Key Laboratory for Behavior and Cognitive Neuroscience, Xi'an, Shaanxi, China

Dr. Zhao is Professor of Psychology at Shaanxi Normal University in China and a principal investigator affiliated with the Key Laboratory for Behavior and Cognitive Neuroscience of Shaanxi Province. Her current research focuses on children's cognitive and psychiatric disorders. Her laboratory is currently working on various topics but mainly focusing on cognitive, neural, genetic, and environmental determinants of developmental dyslexia, dyscalculia, and psychiatric disabilities in children.

The Dyslexia Foundation and the Extraordinary Brain Series

The Dyslexia Foundation (TDF) was founded by William H. "Will" Baker, in collaboration with notable researchers in dyslexia, in the late 1980s. Funds were provided through the generosity of the Underwood and Baker families to support the establishment of the first Dyslexia Research Laboratory at Beth Israel Hospital, Harvard Medical School, Boston, Massachusetts; the laboratory opened in 1982, with a goal of investigating the neural underpinnings of dyslexia. Baker became the director of research for the Orton Dyslexia Society, and he convened top researchers from cognition, neuroscience, and education in a 1987 meeting at the urging of Dr. Albert Galaburda. That scientific symposium was held in Florence, Italy, under the auspices of the Orton Dyslexia Society, with generous support from Emily Fisher Landau. Ideas were presented and discussed at the symposium, with sufficient time to disagree, identify research challenges, and brainstorm solutions—and the concept of a dyslexia symposium series was born. The National Dyslexia Research Foundation (later renamed TDF) was formed in 1989 to focus more specifically on research while the Orton Dyslexia Society continued its primary focus on treatment and education. The new foundation sponsored the next symposium in Barcelona, Spain, in 1990. The Extraordinary Brain Series was born during this second symposium, which was the first to be held under the foundation's auspices!

This volume celebrates the 15th symposium in the Extraordinary Brain Series. Currently, these symposia result in volumes that reflect the papers presented and the discussion that was spurred by those presentations. The series volumes make the thoughts of scholars across various disciplines accessible to all researchers as they tackle various aspects of the behavior, neurobiology, and genetics of dyslexia and learning to read and write. Following is a listing of TDF symposia and the related volumes to date:

I. June 1987, Florence, Italy. Symposium Director: Albert M. Galaburda.

Galaburda, A. M. (Ed.). (1989). *From reading to neurons.* Cambridge, MA: Bradford Books/the MIT Press.

II. June 1990, Barcelona, Spain. Symposium Director: Albert M. Galaburda.

Galaburda, A. M. (Ed.). (1993). *Dyslexia and development: Neurobiological aspects of extra-ordinary brains.* Cambridge, MA: Bradford Books/Harvard University Press.

III. June 1992, Santa Fe, NM. Symposium Director: Paula Tallal.

Chase, C., Rosen, G., & Sherman, G. F. (Eds.). (1996). *Developmental dyslexia: Neural, cognitive, and genetic mechanisms.* Baltimore, MD: York Press.

IV. June 1994, Kauai, HI. Symposium Director: Benita Blachman.

Blachman, B. R. (Ed.). (1997). *Foundations of reading acquisition and dyslexia: Implications for early intervention.* Mahwah, NJ: Lawrence Erlbaum Associates.

V. June 1998, Kona, HI. Symposium Director: Drake Duane.

Duane, D. (Ed.). (1999). *Reading and attention disorders: Neurobiological correlates.* Baltimore, MD: York Press.

VI. June 2000, Crete, Greece. Symposium Director: Maryanne Wolf.

Wolf, M. (Ed.). (2001). *Dyslexia, fluency, and the brain.* Baltimore, MD: York Press.

VII. June 2002, Kona, HI. Symposium Director: Barbara Foorman.

Foorman, B. (Ed.). (2003). *Preventing and remediating reading difficulties: Bringing science to scale.* Baltimore, MD: York Press.

VIII. October 2002, Johannesburg, South Africa. Symposium Director: Frank Wood.

Multilingualism and dyslexia. No publication.

IX. June 2004, Como, Italy. Symposium Director: Glenn Rosen.

Rosen, G. (Ed.). (2006). *The dyslexic brain: New pathways in neuroscience discovery.* Mahwah, NJ: Lawrence Erlbaum Associates.

X. June 2007, Campos do Jordão, Brazil. Symposium Directors: Ken Pugh and Peggy McCardle.

Pugh, K., & McCardle, P. (Eds.). (2009). *How children learn to read: Current issues and new directions in the integration of cognition, neurobiology and genetics of reading and dyslexia research and practice.* New York, NY: Psychology Press, Taylor & Francis Group.

XI. January 2010, Taipei, Taiwan. Symposium Directors: Peggy McCardle, Ovid Tseng, Jun Ren Lee, and Brett Miller.

McCardle, P., Miller, B., Lee, J. R., & Tseng, O. J. L. (Eds.). (2011). *Dyslexia across languages: Orthography and the brain-gene-behavior link.* Baltimore, MD: Paul H. Brookes Publishing Co.

XII. June 2010, Cong, Ireland. Symposium Directors: April Benasich and Holly Fitch.

Benasich, A., & Fitch, H. (Eds.). (2012). *Developmental dyslexia: Early precursors, neurobehavioral markers, and biological substrates.* Baltimore, MD: Paul H. Brookes Publishing Co.

XIII. June 2012, Talinn, Estonia. Symposium Directors: Brett Miller and Laurie Cutting.

Miller, B., Cutting, L. E., & McCardle, P. (Eds.). (2013). *Unraveling reading comprehension: Behavioral, neurobiological and genetic components.* Baltimore, MD: Paul H. Brookes Publishing Co.

XIV. June 2014. Horta, Faial Island, The Azores. Symposium Directors: Carol Connor and Peggy McCardle.

Connor, C. M., & McCardle, P. (Eds.). (2015). *Advances in reading intervention: Research to practice to research.* Baltimore, MD: Paul H. Brookes Publishing Co.

XV. June 2016, Saint Croix, U.S. Virgin Islands. Symposium Directors: Albert Galaburda, Fumiko Hoeft, and Nadine Gaab.

Galaburda, A. M., Gaab, N., Hoeft, F., & McCardle, P. (Eds.). (2017). *Dyslexia and neuroscience: The Geschwind-Galaburda Hypothesis 30 years later*. Baltimore, MD: Paul H. Brookes Publishing Co.

Preface

More than three decades ago, Norman Geschwind and Peter Behan reported an increased prevalence of left-handedness and autoimmune disorders in individuals and families with developmental dyslexia. Following that report, Geschwind collaborated with Albert Galaburda and wrote a three-part paper in the *Archives of Neurology* (1985a, b, c) that discussed developmentally relevant associations among brain development, hormones, immune activity, and brain lateralization, which resulted in human diversity in talents and disabilities. The Dyslexia Foundation hosted a symposium in 2016 that was organized by Galaburda, along with Nadine Gaab and Fumiko Hoeft, to review some of the Geschwind-Galaburda Hypothesis (GGH) claims and predictions and assess the current state of the science regarding them. This volume is based on the presentations and discussions at that symposium.

Because much of the information presented and discussed is at times highly technical, the co-editors of this volume asked the contributing authors to include a summary that includes less technical language at the beginning of each chapter. These summaries are intended to make the volume more accessible to a wider audience because they outline the information that will be presented in each chapter and commentary. In addition, an integrative summary for each of the sections integrates thoughts across each chapter of the section.

The design of the volume itself is also intended to make the material more accessible. In Section I, Galaburda (Chapter 1) introduces the GGH and Urion (Chapter 2) reflects on its impact on clinical practice in child neurology and developmental neuroscience since the mid-1980s, with implications for the future. The following five major sections examine specific areas of scientific investigation responding to and influenced by the GGH, with each chapter making suggestions for future research based on what has been learned since Geschwind and Galaburda's publication of those three seminal papers.

In Section II, four chapters and an integrative summary address brain development, hormonal influences, and immunological issues as interconnected issues or processes. Rubenstein (Chapter 3) examines developmental, genetic, hormonal, and random processes that may play a role in the development of language processing circuits and, thus, may influence or underlie some developmental disorders, including dyslexia. He outlines four specific additional mechanisms regulating brain development that scientists should consider as they explore potential underlying mechanisms of dyslexia. De Vries (Chapter 4) discusses the GGH and sex differences from a whole body perspective and offers new thoughts on the influences of sex differences in the brain, circulating sex hormones during postnatal through adult development, environmental context, and vulnerabilities for functional and behavior disorders. O'Connor, Barrett, and Sefair

(Chapter 5) examine what is known about prenatal maternal distress and its association with a range of neurodevelopmental outcomes in children and discuss these in the context of the GGH, noting a striking confluence with implications for future research. Nguyen (Chapter 6) presents the biological processes of brain development and increasing sex hormone production in middle childhood; she addresses the role of androgens in the pattern of brain maturation and the link with developmental cognitive and behavior changes. The final chapter in this section is the integrative summary by McCarthy. She reflects on the central goal of revealing the origins of brain disorders and pulls together thoughts on developmental impacts, from prenatal through adolescent development, and highlights the plastic and dynamic aspects of brain development, citing the GGH as a framework for such revelation.

Section III's four chapters and an integrative summary focus specifically on sex differences. Ramus, Altarelli, Jednoróg, Zhao, and di Covella (Chapter 7) review their own team's studies on the neuroanatomy of dyslexia, which show differences in boys and girls. The authors assert that the key to understanding dyslexia's neuroanatomical basis lies in brain asymmetry, which differs at least, in part, by sex. Luders (Chapter 8) also examines sex differences in brain anatomy, with a focus on brain size differences, and demonstrates how matching males and females for gross brain size may serve to unscramble whether sex are artefactual or exist independently of size effects. Hong (Chapter 9) focuses on sex differences in learning disorders, offering the model of sex chromosome anomalies such as those found in Turner and Klinefelter syndromes. He calls for future work to further explore the mounting evidence that sex plays a role in learning disorders, including dyslexia. Fitch, Bodge, Szalkowski, and Smith (Chapter 10) present findings on sex differences in atypical cognitive outcomes in animal models that are relevant to language and reading and bring these to bear on the issue of greater prevalence of dyslexia among males. Cutting's integrative summary notes the lack of clarity of the relation between sex and dyslexia diagnosis as we look across neuroimaging, genetics, and animal model research linked to the GGH, and it offers suggestions for future directions that research might take to clarify this picture.

Section IV on laterality contains only two chapters and an integrative summary. Grigorenko (Chapter 11) presents information on genetic concepts—mosaicism, epigenetic, and the endophenotypes—and how using these can help us understand complex human traits such as laterality and its underlying processes by breaking complexity down into component parts. Ocklenburg and Peterburs (Chapter 12) focus on genetic factors underlying handedness and left-hemisphere language dominance; they argue that the assumptions of the GGH offer an opportunity for development of a multifactorial model that could integrate neurogenetic evidence into those core constructs. Paracchini's integrative summary points out that complex processes must be controlled by more than simple genetic models and embracing advances in genetic sequencing will likely be important in advancing our understanding of laterality.

Reading and dyslexia are specifically addressed with regard to genes and behavior in Section V. Kornilov and Grigorenko (Chapter 13) begin this section with an overview of genetics and reading disability, including genome-wide association studies, studies using neuroimaging phenotypes, and those examining the role of common genetic variants. They argue strongly for the importance of studies that integrate all three in the study of reading disability. Hoeft and Hancock (Chapter 14) describe their work on intergenerational transmission of reading and the brain networks that underlie it. These authors introduce the intergenerational neuroimaging approach to studying the relation of cognitive and neural phenotypes across generations from parent to child; they hope this process will affect early identification and intervention and result in the identification of modifiable intervention targets. Gaab, Yu, and Ozernov-Palchik (Chapter 15) also examine the heritability of dyslexia and early identification and intervention, combining neuroimaging of infants and young children with behavioral research to identify children at risk. Feldman (Chapter 16) examines a different risk factor (i.e., preterm birth) and compares white matter brain differences and their correlations with reading and language skills in children born pre- and full-term. Her findings suggest a loss of hemispheric specialization in the preterm group, and she calls for additional studies to explore the GGH and determine these children's specific reading intervention needs. Landi's integrative summary notes the value of being able to more accurately identify the relative contributions of genes and environment in the atypical development of reading brain regions, but it also notes that the advanced methods that have been discussed come with an increased need for specialized training and with a requirement to communicate to educators and policy makers both the importance and the nuances of the findings these tools enable us to obtain.

These section chapters and integrative summaries explore key elements of the GGH, looking at how far we have come and how many questions have been answered. They also look, however, at how we might build on both the original GGH and the recent work that addresses many of the issues it raised. They suggest how the field might move forward with modern technologies and advanced methods to continue to tease apart the puzzles of the developing brain, learn what underlies differences in brain development, how genes and environment interact with our neurocircuitry and neurochemistry in ways that result in individuals who read well or who have difficulty learning to read, and how best to prevent or remediate those difficulties. The contributing authors have attempted to communicate highly technical information in ways that are accessible to an audience broader than just scientific colleagues, and the editors have attempted to design a book that also facilitates information sharing for a broader audience. The information gained since the mid-1980s and the directions recommended for the future are important and clearly worth sharing not only with researchers but also with practitioners, parents, and all those interested in dyslexia, its origins, and its prevention and remediation.

REFERENCES

Geschwind, N., & Galaburda, A. M. (1985a). Cerebral lateralization. Biological mechanisms, associations, and pathology: I. A hypothesis and a program for research. *Archives of Neurology, 42*(5), 428–459.

Geschwind, N., & Galaburda, A. M. (1985b). Cerebral lateralization. Biological mechanisms, associations, and pathology: II. A hypothesis and a program for research. *Archives of Neurology, 42*(6), 521–552.

Geschwind, N., & Galaburda, A. M. (1985c). Cerebral lateralization. Biological mechanisms, associations, and pathology: III. A hypothesis and a program for research. *Archives of Neurology, 42*(7), 634–654.

Acknowledgments

This volume is the result of a great effort by many individuals. First, it would have been impossible without The Dyslexia Foundation (TDF), sponsor of the symposium that gave rise to it. We owe a huge thanks to TDF founder and president Will Baker and Amelia Baker Lauderdale, along with several of TDF's trustees and scientific advisory board members. Without them there would be no Extraordinary Brain Symposia (EBS) and, thus, no EBS series of volumes!

We also must acknowledge the authors who traveled to Saint Croix, U.S. Virgin Islands to present their own work and thoughts, engaged in deep discussions with symposium participants, and wrote and revised chapters for this volume, often meeting stringent deadlines. In addition, we acknowledge the practitioners (teachers, educators, administrators, and clinicians), researchers, graduate students, and postdoctoral fellows who also traveled and not only listened attentively and thought deeply but also discussed, questioned, and commented, helping to inform this volume.

We and TDF offer our heartfelt thanks to the following individuals, schools, and businesses that generously supported EBS XV: AIM Academy, Athena Academy, the Brehm School, the Carroll School, Cerebral Matters, the Chartwell School, Curry Ingram Academy, the Eagle Hill Southport School, the Howard School, Joan McNichols, the Summit School, the Westmark School, Wilson Language Training, and five individuals and two foundations that wish to remain anonymous. In addition, special thanks go to the officials and individuals from St. Croix, U.S. Virgin Islands who helped make the symposium a success. We wish to specifically acknowledge the receptions hosted and supported by The Buccaneer Hotel and the Cruzan Rum Co. and the following people and institutions in the U.S. Virgin Islands for their help with the arrangements for this symposium: Kimberly Mills, PhD, BCBA-D, VIUCEDD (Center for Exceptional Developmental Disabilities); Maureen Moorehead, Virgin Islands Department of Education, Division of Special Education; Kari Loya, Head of School, and Wendy F. Canning, F/AOGPE, both of the Good Hope Country Day School, St. Croix, U.S. Virgin Islands.

Finally, thanks go to the flexible, helpful, and extraordinarily capable editorial staff at Paul H. Brookes Publishing Co., especially Astrid Pohl Zuckerman (who was an immense help in guiding the authors on what to expect and helping us to think through how best to design this volume). Astrid and the Brookes staff supported us and ensured a high-quality volume. Thank you for your help and patience on this book. We loved working with you!

Co-editors,
Al Galaburda, Nadine Gaab, Fumiko Hoeft, and Peggy McCardle

SECTION I

The Geschwind-Galaburda Hypothesis and Dyslexia

CHAPTER 1

The Geschwind-Galaburda Hypothesis

Albert M. Galaburda

SUMMARY

The Geschwind-Galaburda Hypothesis (GGH), which was first published as three consecutive papers (1985a, b, and c), is, despite its name, a theory comprising several testable hypotheses that address associations among brain development and function, sex hormones, immunological health, and brain lateralization. Clinical manifestations of this theory may lead to individuals with special talents or disabilities. The theory was inspired by the report of Geschwind and Behan (1982), indicating a relationship among left-handedness, developmental dyslexia, and immune disorders.

The GGH suggested that testosterone in utero slowed development of the left hemisphere, increasing the risk for left-handedness and language-based learning disorders, as well as superior development of right hemisphere skills, both of which would occur more often in boys. Testosterone also suppressed immune function, leading to an excess of immune-based disorders. An elaboration of the theory posited that the retardation of left-hemisphere development could lead to neuronal migration anomalies, which represented the mechanism underlying the language-based learning disorders. The neuronal migration anomalies were recreated in rodents by using various approaches, including early, timed brain injury and knock-down of dyslexia risk genes, but not simply by the effects of testosterone, although the latter did alter the functional consequences of the developmental anomalies. Additional work in genetically modified mice (Truong et al., 2014; Wang et al., 2011) showed that structural anomalies could, but were not necessary to produce cortical dysfunction. In addition, recent work on the effects of cilia function (Hamada, Meno, Watanabe, & Saijoh, 2002) on organ laterality were suggested to play a role in the anomalous brain lateralization seen in dyslexia, especially as cilia dysfunction is associated with some dyslexia risk genes. The GGH generated a great deal of research, which continues to this date.

INTRODUCTION: THE ORIGINS OF THE GESCHWIND-GALABURDA HYPOTHESIS

This chapter reflects a presentation and subsequent discussions that took place during a conference in June 2016, which was organized and sponsored by The Dyslexia Foundation. Both the presentation and this chapter rely heavily on the three-part paper published by Geschwind and Galaburda (1985a, b, c).

The GGH is an attempt to relate observations not only about a specific learning disorder (i.e., dyslexia) but also other male-predominant developmental disorders (e.g., stuttering, autism spectrum disorder, Tourette syndrome) to cerebral lateralization for a variety of cognitive tasks, including handedness and hemispheric dominance for language, certain autoimmune disorders, and hormonal and genetic characteristics. The papers specifically stressed male–female differences in the incidence and occurrence of these phenomena and attributed the male sex steroid testosterone, or its metabolites, as a main causative role, rather than the environment or direct genetic influences. The hypothesis states that intrauterine testosterone increases anomalous functional or physiological lateralization (left-handedness, ambidexterity, and right-hemisphere or bilateral language in the brain), anatomical brain asymmetry (affecting the language-relevant planum temporale [PT]), learning disabilities (dyslexia, stuttering, Tourette syndrome, autism spectrum disorder), learning superiority (architects, visual artists), and immunological and other disorders (asthma, thyroid disorders, ulcerative colitis, premature graying of the hair).

The GGH is not really a hypothesis but a theory that is complex enough to generate several testable hypotheses on sex differences, handedness, and hemispheric dominance, specifically regarding the relations and interactions among brain development, hormones, immune activity, and brain lateralization, which resulted in human diversity in talents and disabilities. The work of Peter O. Behan, a neuroimmunologist from Glasgow, Scotland, initially contributed to the GGH; thus, the theory has also been called the Geschwind-Behan-Galaburda Hypothesis. Most important, however, is that the theory is primarily the brainchild of Norman Geschwind (see Figure 1.1; see the dedication in this volume).

The Planum Temporale and Cerebral Asymmetry

Although Geschwind died in 1984, only months before the publication of the GGH, it is likely that kernels of the theory had their origin as early as the mid-1960s, when he and Levitsky (1968) revisited a set of findings by Von Economo and Horn (1930) and Pfeifer (1936). These findings suggested that the PT, the triangle-shaped cortical region lying immediately posterior to the transverse gyrus of Heschl containing the primary auditory cortex, was larger on the left-side.

Geschwind and Levitsky (1968) were stimulated by these anecdotal reports and dissected 100 human brains. They determined that the left planum was larger in 65% of the brains, it was larger on the right in 11%, and it was of equal size on the two sides in the remaining 24%. Geschwind became curious about the fundamental causes of cerebral asymmetry and its

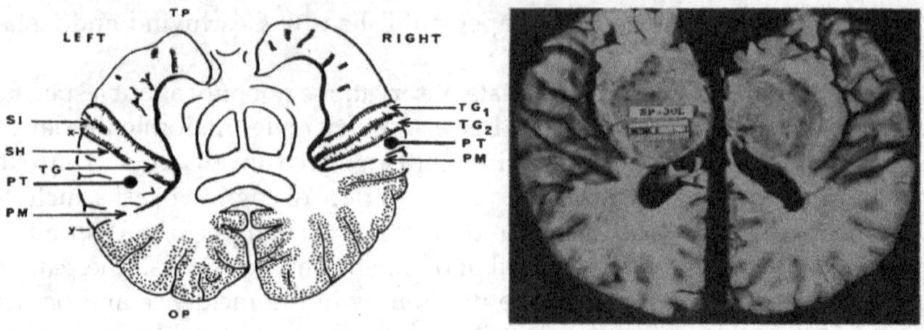

Figure 1.1. Original figure from Geschwind and Levitsky (1968) showing a larger right planum temporale (PT) and a second observation, the more common duplication of Heschl's gyrus or transverse gyrus on the right side. From Geschwind, N., & Levitsky, W. (1968). Human brain: Left-right asymmetries in temporal speech region. *Science, 161*(3837): 186–187; reproduced with permission.

biological associations. In 1976, he encouraged Albert Galaburda, a physician in the last year of neurology residency training, to look deeper for the underlying causes of brain asymmetry. But anatomical brain asymmetry was only part of the investigation of brain functioning. Links between this asymmetry and behaviors, such as handedness and learning, and to other conditions involving the immune system were also of interest.

Left-Handedness, Language-Based Learning Disorders, and Immunological Conditions

Geschwind was a dedicated and keen clinical observer. He used to say that most of his ideas came from listening to and observing his patients. An informal cataloguing of conversations and observations led him to two conclusions.

1. Unusual manifestations of acquired aphasia (an impairment of language comprehension or production usually caused by injury to the left-hemisphere of the brain) and various forms of language-based LDs occurred more frequently in left handers.
2. His left-handed patients seemed to complain of more ailments than his right-handed patients, especially allergic and autoimmune conditions.

These anecdotal observations led to studies with Behan, first of self-reported immunological conditions and then of a hospital record-confirmed assessment of left-handed people arriving at a shop for left-handed objects in London (Geschwind & Behan, 1982). Both indicated more autoimmune conditions and language-based LDs in left handers and their family members.

Sex Differences, Handedness, and Hemispheric Dominance: The Testosterone Hypothesis

Sex differences is the fourth component of the theory. LDs that affected left-hemisphere brain functions (e.g., dyslexia, stuttering, autism spectrum disorder) were far more common in boys than in girls. In addition, the affected individuals often exhibited superior performance in right-hemisphere functions (e.g., visuospatial abilities). These gender differences could have been the result of X or Y chromosome factors. If the differences were linked to the X chromosome, however, then the male predominance would be very high—much higher than that observed—or if the differences were linked to the Y chromosome, then these disorders would be nonexistent in females. Thus, Geschwind posited that the excess of left-handedness and LDs in males suggested a failure of proper development of left-hemisphere dominance patterns that were testosterone mediated; this was possible because males and females are exposed to intrauterine testosterone, albeit at different concentrations. This was the origin of the testosterone hypothesis of the GGH, and the testosterone hypothesis was at the core of the theory. Summing up, LDs linked to left-hemisphere function are noticeably more prevalent in males, but they also occur in females, which suggests that testosterone exposure in utero mediates the development of proper patterns of left-hemisphere dominance. The fact that male fetuses are exposed to greater concentrations of testosterone may be linked to the greater prevalence of these LDs among males.

ELEMENTS OF THE GESCHWIND-GALABURDA HYPOTHESIS

The GGH represented an attempt to link observations on left-handedness, brain asymmetry, learning differences, immune disorders, and male–female differences. Testosterone effect (a combination of testosterone level, the distribution of testosterone receptors, and the sensitivity and specific activity at these receptors) is the unifying causal factor that occurs during development. In brief, the theory postulated that testosterone acting in utero slowed the development of the left hemisphere of the human cerebrum, which, in turn, led to

- Possible enhancement of right-hemisphere development and more left-handedness (or underdeveloped right-handedness)
- Possible LDs involving left-hemisphere functions
- Possible superior skills in right-hemisphere functions
- Greater prevalence in males
- A disturbance in the development of the immune system by the direct action of testosterone on the thymus gland and other immunological targets that are poorly understood

Related Research: Gross Cerebral Asymmetries

The theory depended heavily on the finding of structural brain asymmetries. Several early observers (Flechsig, 1876; Pfeifer, 1936; Von Economo & Horn, 1930) reported on PT asymmetries as well as others (e.g., asymmetry of the crossing of the motor tracts in the brain stem on their way to the spinal cord). LeMay (1976) examined human brain endocasts that were 30,000 years old and reported asymmetries in Neanderthals. She also examined cerebral arteriograms of adult humans and found asymmetries in the sylvian point, the spot where the middle cerebral artery exits to the surface of the brain posteriorly (LeMay & Culebras, 1972). The main observations were that the sylvian fissure was longer on the left and the sylvian point was higher on the right side than on the left side. It was shown later that the length asymmetry reflected the longer PT on that side.

It became important to find the cause(s) of these asymmetries. The right shift theory is a genetic explanation proposed by Annett (1964) that suggests the presence of a laterality gene causes the brain to acquire the standard asymmetry, causing the individual to become right handed. The absence of this laterality, or right shift gene, produced an individual who would be randomly right or left handed 50% of the time. Annett even calculated the prevalence of the gene in the population to explain the numbers of right and left handers. Annett could not distinguish between right handers having the gene and right handers arising from random assignment, but no left handers possessed the gene. The GGH, however, predicted that even in the presence of the right shift gene, testosterone in utero could affect the degree of expression of this gene in an epigenetic fashion and impede the full shift to the right-handed pattern, leading to incomplete left-greater-than-right asymmetry, or even asymmetry in the opposite direction, as well as LDs. The tension between the Annett theory and the GGH is that the former did not require any nongenetic factors, whereas the latter introduced a factor in the form of testosterone, which acted as an epigenetic regulator of the gene effect. As of this writing, the Annett theory, which has in practice been replaced or expanded on by the new discoveries on cilia function—which responds to multiple genetic and epigenetic actions (Hamada, Meno, Watanabe, & Saijoh, 2002)—has not included any additional testosterone effect. This said, without hormonal influences (see Chapter 6), neither the Annett theory nor the cilia biology can explain the male predominance that inspired the GGH.

Related Research: Architectonic Cerebral Asymmetries

Galaburda and colleagues (Galaburda & Sanides, 1980; Galaburda, Sanides, & Geschwind, 1978) published a series of papers attempting to find the fundamental underpinnings of brain asymmetries. First, they found that asymmetries in gross anatomy reflect asymmetries in the volume (surface area x cortical thickness) of architectonic areas (cf. Brodmann areas), and asymmetries in the PT are specifically accounted for by asymmetries in the auditory temporoparietal association area on the posterior upper surface of the temporal lobe. Other architectonic areas comprising the perisylvian language zones (e.g., Brodmann area 44 and 39) also tended to favor the left side (Eidelberg & Galaburda, 1984; Galaburda, 1980; Galaburda & Geschwind, 1980). The GGH attempted to provide explanations for these language-related asymmetries and for the cases in which the asymmetries develop anomalously, and here are some of the findings of the architectonic work that followed the publication of the GGH but did not support the GGH. This work was carried out on an adult rodent cerebral cortex because technical aspects for the preparation of tissue could be better controlled for precise quantitative assessment (Galaburda, Aboitiz, Rosen, & Sherman, 1986).

Thus, one claim of the GGH in the case of anomalous asymmetry (i.e., less left-sided predominance) was that as the left hemisphere was inhibited during development by the effect of testosterone, the right hemisphere grew larger than normal, shifting brain structures from the left to the right. It was as if there was a given amount of brain that was variably distributed more to the left, more symmetrically, or even more to the right, according to the magnitude of the testosterone effect. In the case of language areas, correcting for variability of general brain size, everyone was supposed to have an equivalent amount, and the only thing that changed was how this equivalent amount was stored in the left and right hemispheres. Thus, correcting for the overall variation in brain size, the GGH predicted that the left planum area plus the right planum area was constant (Galaburda, Rosen, & Sherman, 1990). In fact, this was not the case when asymmetric and symmetric areas were compared. The total amount of tissue in the area did not remain constant in human and animal studies, in which there is variation in degree of asymmetry of a given gross or architectonically defined area. Instead, it varied inversely with the degree of asymmetry—the more asymmetric the areas were between the two sides of the brain, the smaller their sum. Thus, symmetric areas were larger than asymmetric areas, which suggested that asymmetry could represent a pruned-down version of the symmetric case and not simply a different distribution. Furthermore, at the architectonic level, symmetric brain areas in one brain contained a larger complement of neurons (as opposed to larger neurons) than homologous asymmetric brain areas in another brain (see Figure 1.2) (Galaburda et al., 1986; Rosen, Sherman, & Galaburda, 1991, 1992).

Several additional studies showed that the lateral difference in the number of neurons was neither the result of differential neuronal proliferation prior to neuronal migration, nor differential cell death or pruning after neuronal migration (Rosen et al., 1991). Although this was not looked at directly, it was likely that asymmetry in the number of neurons on the two sides reflected early patterning events before neuroblast proliferation, neuronal migration, or neuronal death and maturation (see Chapter 3). If there was pruning at all, then it had to occur very early in the life of the embryo. The GGH considered that a mechanism for shifted dominance to the right hemisphere by testosterone implicated greater cell death on the left, with compensatory lesser cell death on the right, and this was implicitly

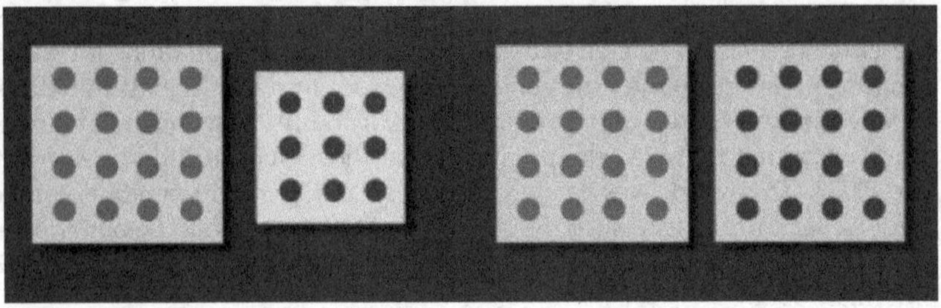

Figure 1.2. The generation of asymmetric brain areas requires developmental reduction from an originally symmetric state. Indirect evidence suggests that this reduction occurs before neuronal proliferation and neuronal migration to the incipient cerebral cortex and may be mediated by cilia function.

considered to be a postmigrational event. Instead, the likely scenario is that initial patterning of the cortex is symmetric, and early events, perhaps under the effect of cilia and possibly still modulated at least in part by testosterone, produce reduction of one side—the right side in the case of language areas—so that fewer neuroblasts remain available on the right side and a smaller architectonic area is subsequently generated.

Related Research on Language-Based (Left-Hemisphere) Learning Disabilities: Cortical Malformations and Left-Handedness

The GGH made a strong claim about the origin of left-hemisphere learning disabilities such as dyslexia, autism spectrum disorder, and stuttering. Thus, implicating testosterone again, another hypothesis of the theory was that retardation of growth of the left hemisphere was so severe in some cases that it caused neuronal migration anomalies on that side, which, in turn, caused the LDs, either alone or in combination with decreased leftward asymmetry of the language areas and left-handedness. No attempt was made to distinguish cases in which dyslexia versus stuttering versus autism would result, but the implicit assumption at that time was that testosterone acted on different genetic backgrounds to produce different disorders. Suffice it to say, testosterone effects occurred in all three conditions, thus explaining the male prevalence in all three. Subsequent experimental research failed again to support the notion that testosterone was responsible for neuronal migration anomalies, although another, previously unsuspected action of the male hormone probably did play a role in the gender difference.

Cortical Malformations Galaburda and Kemper (1979) reported the case of a dyslexic young man whose brain showed neuronal migration anomalies. Multiple instances of molecular layer heterotopia were present in the perisylvian regions of both hemispheres, but they were substantially more prominent on the left, as well as one focus of micropolygyria affecting the auditory regions of the left temporal lobe. Similar small cortical malformations were present in several other dyslexic brains later studied at autopsy (Galaburda, Sherman, Rosen, Aboitiz, & Geschwind, 1985). The first strike against the GGH suggesting that neuronal migration anomalies were the result of excessive inhibition of left-hemisphere development was the fact that neuronal migration abnormalities were found on both sides, even if more commonly on the left. Although the theory hypothesized an inhibitory effect on the left, it also claimed a stimulatory effect on the right. Although there was the possibility that neuronal migration anomalies could result from over-inhibition or over-stimulation and would be worse with inhibition than with stimulation, no data were ever gathered to support this possibility. In the end, the pathogenesis of the malformations was related to different factors such as injury, infection, toxic exposure, or genetic defects (Kuzniecky & Barkovich, 1996; Pilz, Stoodley, & Golden, 2002), without evidence that testosterone has any role to play in neuronal migration abnormalities.

Neuronal heterotopias in the molecular layer of the cortex and micropolygyria (both found in the dyslexic brain) have been induced in rodent brains by applying a freezing probe on the skull of the newborn animal during a time when neuronal migration to the cerebral cortex is continuing. Neuronal migration anomalies can only occur if the causative agent acts during the time neuronal migration is actively proceeding to the cerebral cortex and not later (see Figure 1.3) (Humphreys, Rosen, Press, Sherman, & Galaburda, 1991). Different, but perhaps associated, forms of neuronal migration anomalies have also been caused in the fetal rat brain exposed to short hairpin micro ribonucleic acid sequences constructed against genes in the rat homologous to human dyslexia risk genes (Rosen et al., 2007; Szalkowski et al., 2012; Wang et al., 2006) and, in some cases, in mice knocked out for these genes (Truong et al., 2014; Wang et al., 2011). These manipulations were identical in males and females, and the cortical effects were also the same, but there was an effect of testosterone. Thus, males with induced cortical malformations showed secondary changes in the thalamus, whereas females did not. But the secondary thalamic changes emerged in female rats when they were exposed perinatally to testosterone (Rosen, Herman, & Galaburda, 1999). Thus, testosterone did not influence cortical malformations per se during the experiment on the rodent, but rather a form of secondary thalamic plasticity. An additional interesting finding was that induction of cortical malformation resulted in auditory processing deficits in male animals but not female animals, suggesting that it was the response to the initial manipulation that produced the deficits, rather than the cortical malformations themselves. The human thalamus is asymmetric (Eidelberg & Galaburda, 1982); thalamic anomalies have also been reported in the human dyslexic brain (Galaburda &

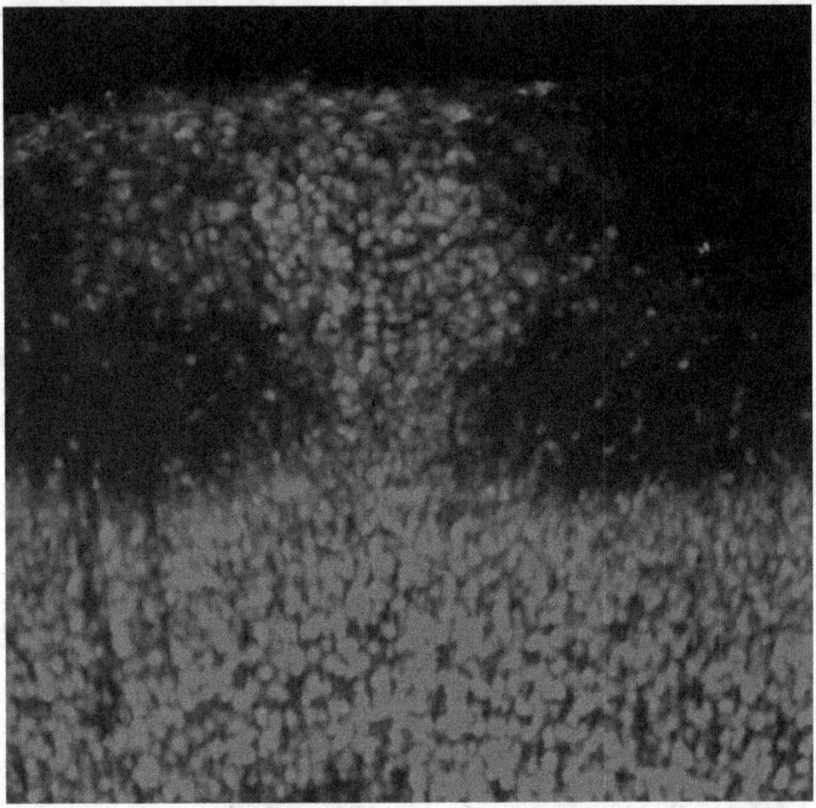

Figure 1.3. Molecular layer ectopia in the rat cortex. Neurons are stained with red-fluorescent protein. The cloud of neurons lying above in the photograph have migrated abnormally beyond their standard locations in the subjacent cortical layers. These malformations occur spontaneously in the dyslexic cerebral cortex and can be induced in rodent brains for further analysis.

Eidelberg, 1982). It is possible that the action of the human cortical anomalies in dyslexia have been overestimated and the effects of the secondary thalamic changes have been underestimated. Effects of testosterone on injury-related plasticity were not something considered in the GGH, but they appear to account for the sex differences and should be further studied in the biology of left-hemisphere learning disabilities. It seems to account for sex differences not by causing neuronal migration anomalies per se, but rather by modifying plasticity in response to a negative developmental event.

Left-Handedness As previously noted, the biology of cerebrocortical abnormalities in relation to testosterone seems to be more complex than simply slowing the development of the left hemisphere. We do not understand the relationship of testosterone to handedness. We also do not know why anomalous hand dominance is more common in males, and the answer may have something to do with testosterone; if it does, then it may be by mechanisms other than slowing the development of the left hemisphere. The biology of cilia has come to the forefront in the study of lateralization (Hamada et al., 2002). Cilia seem to play an important role in the lateralization of somatic organs, and the absence of cilia function may lead to situs inversus (Pennekamp, Menchen, Dworniczak, & Hamada, 2015). Interestingly, 50% of animals with abnormal cilia develop a heart pointing to the right, and 50% develop a heart pointing to the left. The same random lateralization is true for the positioning of the liver and spleen.

In other words, the absence of cilia function leads to random lateralization of the body organs, and this is strikingly similar to what Annett's (1964) right shift theory predicts; thus, the absence of the right shift gene leads to random lateralization in the brain. Annett's lateralization gene may not be a single gene but may instead reflect dysfunction of any of the genes that govern normal cilia function. This statement makes a strong assumption that cannot be adequately defended at this time and implicates cilia dysfunction in brain lateralization as well as in body organ lateralization. Furthermore, Annett's theory makes additional predictions. For instance, the cilia and Annett's explanations predict that in any group of right handers there will be some who are anomalous right handers because they randomly became right handed from lack of the normal drive, either cilia or gene based. An equal number of left handers will be anomalous for the same reason. Implications of this biology of handedness would be that there are so-called normal right handers and anomalous right handers based on whether they have the right shift gene or cilia effect, and all left handers are anomalous because all lack the gene or cilia. Is this an obligatory conclusion about left handers? Not necessarily! Thus, for instance, there may be drives equivalent to the right shift drives that make people left handed, and some within this group may also lack this left shift drive. In this newly postulated left shift theory, the presence of left shift factors would produce standard left handers and the absence would lead to random left and right handers who are anomalous because they are missing the left shift effect. The frequency in the population of this postulated left shift factor must be lower than that of the right shift factor; otherwise, there should be an equal number of standard right and left handers, which is not the case. Whether it is from the absence of the right shift factor or the absence of the postulated left shift factor, there should be an equal number of anomalous right- and left-handed individuals, albeit a greater number of standard right handers than left handers. Furthermore, it is not currently possible to distinguish anomalous from standard right handers and left handers; this fact would tend to produce problems for studies attempting to uncover relationships between handedness and other traits, including disorders such as learning disabilities and immunological disorders, because anomalous right and left handers would be mixed in with the standard groups.

Left Handedness, Immune Disorders, and Language-Based Learning Disabilities Geschwind and Behan (1982) uncovered a relationship among LDs, immune disorders, and left-handedness, which became fodder for the GGH. Yet, following the publication of the three-part paper (Geschwind & Galaburda, 1985a, b, c), several papers failed to find the same relationship, whereas others confirmed the original findings (e.g., Bryden, McManus, & Bulman-Fleming, 1994; Galaburda, 1990; Hugdahl, Ellertsen, Waaler, & Klove, 1989). Leaving aside the null hypothesis (the failure to demonstrate a finding does not mean that there is no finding), there may be other reasons for this inconsistency. Assume that the absence of the right shift factor (or the postulated left shift factor for that matter; I will call them the laterality factors for the present discussion) is accompanied by an altogether different brain organization and represents an important risk factor for language-based learning disabilities (i.e., a sine qua non). If one examines a population of individuals with these disorders, then nearly 50% of them will be right handed and 50% will be left handed because the absence of the laterality factors produces random handedness. Therefore, studies performed in this way will not find a prevalence of left-handedness in a population of individuals with the mentioned disabilities. If, however, a general population of left handers are sampled, then a greater proportion of them will have a language-based learning disability than a similar population of right handers because anomalous left handers represent a greater proportion of the total number of left handers (standard and anomalous) than anomalous vis-à-vis total number of right handers. Unless attention is paid to the way the study is constructed, there will be disagreement as to whether left-handedness is associated with a language-based learning disability. A similar argument can be made regarding immunological conditions. It is also likely that absence of laterality factors occurs in a very small proportion of the population to produce either anomalous left- or right-handedness; studies need large *n*'s to produce robust, replicable findings. This is, in part, a likely explanation for the lack of agreement on the questions of handedness, immune disorders, and language-based LDs among the papers that followed the publication of the GGH.

CONCLUSION

The GGH postulated that in utero testosterone slows the development of the left hemisphere, thus promoting superior development of the right hemisphere, and interferes with the proper setup of the immune system so that males and testosterone-exposed females will exhibit increased left-handedness, language-based LDs, better right-hemisphere skills, and more immunological disorders. The increased incidence of LDs is supposed to be doubly caused by testosterone-based slowing of development of left hemisphere language areas and malformations caused by inhibited migration of neurons to the cerebral cortex. Several findings from the 1930s through the 1980s provided evidence for building the theory, including excess left-handedness and language-based LDs in boys, the presence of variable handedness in the population, the presence of variable brain asymmetry, the link between left-handedness and language-based LDs, the link between left-handedness and immunologically derived human medical conditions, and the finding of neuronal migration anomalies in the brains of individuals who have dyslexia.

A great deal of research since the publication of the GGH (Geschwind & Galaburda, 1985a, b, c) has shed additional light on the biology of laterality and the mechanisms underlying neuronal migration abnormalities—enough to cast doubt on the role of testosterone, at least the role the GGH attributed to testosterone. Testosterone clearly seems to have a role in developmental plasticity, which may affect handedness, but this remains hypothetical. In fact, the role of testosterone in handedness remains a mystery and will require additional research. Similarly, testosterone does not seem to cause neuronal migration anomalies as proposed in the GGH, but it may modulate the anatomical functional consequences of these neuronal migration anomalies. The role of cilia in the production of organ laterality has come to the forefront of research, but we have yet to learn what cilia have to do with lateralization of the brain, even though they are ubiquitous in the central nervous system. If cilia play a role in handedness formation, then this role will provide a cell-molecular mechanism to explain the observations of Annett (1964) and her right shift theory and will help dovetail some predictions of the GGH. In addition to testosterone's effects on plasticity and recovery of function, it will be useful to investigate whether it plays a role in cilia function and laterality in the brain. Finally, a coherent understanding of the immunological connection to testosterone is beyond the reach of the present discussion.

REFERENCES

Annett, M. (1964). A model of the inheritance of handedness and cerebral dominance. *Nature, 204*, 59–60.

Bryden, M. P., McManus, I. C., & Bulman-Fleming, M. B. (1994). Evaluating the empirical support for the Geschwind-Behan-Galaburda model of cerebral lateralization. *Brain and Cognition, 26*(2), 103–167.

Eidelberg, D., & Galaburda, A. M. (1982). Symmetry and asymmetry in the human posterior thalamus I: Cytoarchitectonic analysis in normal persons. *Archives of Neurology, 39*(6), 325–332.

Eidelberg, D., & Galaburda, A. M. (1984). Inferior parietal lobule: Divergent architectonic asymmetries in the human brain. *Archives of Neurology, 41*(8). 843–852.

Flechsig, P. (1876). *Die leitungsbahnen in gehirn und ruckenmark des menschen auf grundentwicklungs geschlichticher untersuchungen.* Leipzig, Germany: W. Engelmann.

Galaburda, A. M. (1980). Broca's region: Anatomic remarks made a century after the death of its discoverer. *Review of Neurology (Paris), 136*(10), 609–616.

Galaburda, A. M. (1990). The testosterone hypothesis: Assessment since Geschwind and Behan, 1982. *Annals of Dyslexia, 40*(1), 18–38. doi:10.1007/BF02648138

Galaburda, A. M., Aboitiz, F., Rosen, G. D., & Sherman, G. F. (1986). Histological asymmetry in the primary visual cortex of the rat: Implications for mechanisms of cerebral asymmetry. *Cortex, 22*(1), 151–160.

Galaburda, A. M., & Eidelberg, D. (1982). Symmetry and asymmetry in the human posterior thalamus II: Thalamic lesions in a case of developmental dyslexia. *Archives of Neurology, 39*(6), 333–336.

Galaburda, A. M., & Geschwind, N. (1980). The human language areas and cerebral asymmetries. *Revue Médicale Suisse Romande, 100*(2), 119–128.

Galaburda, A. M., & Kemper, T. L. (1979). Cytoarchitectonic abnormalities in developmental dyslexia: A case study. *Annals of Neurology, 6*(2), 94–100. doi:10.1002/ana.410060203

Galaburda, A. M., Rosen, G. D., & Sherman, G. F. (1990). Individual variability in cortical organization: Its relationship to brain laterality and implications to function. *Neuropsychologia, 28*(6), 529–546.

Galaburda, A., & Sanides, F. (1980). Cytoarchitectonic organization of the human auditory cortex. *Journal of Comparative Neurology, 190*(3), 597–610. doi:10.1002/cne.901900312

Galaburda, A. M., Sanides, F., & Geschwind, N. (1978). Human brain: Cytoarchitectonic left-right asymmetries in the temporal speech region. *Archives of Neurology, 35*(12), 812–817.

Galaburda, A. M., Sherman, G. F., Rosen, G. D., Aboitiz, F., & Geschwind, N. (1985). Developmental dyslexia: Four consecutive patients with cortical anomalies. *Annals of Neurology, 18*(2), 222–233. doi:10.1002/ana.410180210

Geschwind, N., & Behan, P. (1982). Left-handedness: Association with immune disease, migraine, and developmental learning disorder. *Proceedings of the National Academy of Science USA, 79*(16), 5097–5100.

Geschwind, N., & Galaburda, A. M. (1985a). Cerebral lateralization. Biological mechanisms, associations, and pathology: I. A hypothesis and a program for research. *Archives of Neurology, 42*(5), 428–459.

Geschwind, N., & Galaburda, A. M. (1985b). Cerebral lateralization. Biological mechanisms, associations, and pathology: II. A hypothesis and a program for research. *Archives of Neurology, 42*(6), 521–552.

Geschwind, N., & Galaburda, A. M. (1985c). Cerebral lateralization. Biological mechanisms, associations, and pathology: III. A hypothesis and a program for research. *Archives of Neurology, 42*(7), 634–654.

Geschwind, N., & Levitsky, W. (1968). Human brain: Left-right asymmetries in temporal speech region. *Science, 161*(3837), 186–187.

Hamada, H., Meno, C., Watanabe, D., & Saijoh, Y. (2002). Establishment of vertebrate left-right asymmetry. *Nature Reviews Genetics, 3*(2), 103–113. doi:10.1038/nrg732

Hugdahl, K., Ellertsen, B., Waaler, P. E., & Klove, H. (1989). Left and right-handed dyslexic boys: An empirical test of some assumptions of the Geschwind-Behan hypothesis. *Neuropsychologia, 27*(2), 223–231.

Humphreys, P., Rosen, G. D., Press, D. M., Sherman, G. F., & Galaburda, A. M. (1991). Freezing lesions of the developing rat brain: A model for cerebrocortical microgyria. *Journal of Neuropathology and Experimental Neurology, 50*(2), 145–160.

Kuzniecky, R. I., & Barkovich, A. J. (1996). Pathogenesis and pathology of focal malformations of cortical development and epilepsy. *Journal of Clinical Neurophysiology, 13*(6), 468–480.

LeMay, M. (1976). Morphological cerebral asymmetries of modern man, fossil man, and nonhuman primate. *Annals of the New York Academy of Science, 280*, 349–366.

LeMay, M., & Culebras, A. (1972). Human brain: Morphologic differences in the hemispheres demonstrable by carotid arteriography. *New England Journal of Medicine, 287*(4), 168–170. doi:10.1056/NEJM197207272870404

Pennekamp, P., Menchen, T., Dworniczak, B., & Hamada, H. (2015). Situs inversus and ciliary abnormalities: 20 years later, what is the connection? *Cilia, 4*(1), 1–12. doi:10.1186/s13630-014-0010-9

Pfeifer, R. A. (1936). *Pathologie der Horstrahlung und der corticaler Horsphäre* (Vol. 6). Berlin, Germany: Springer Verlag.

Pilz, D., Stoodley, N., & Golden, J. A. (2002). Neuronal migration, cerebral cortical development, and cerebral cortical anomalies. *Journal of Neuropathology and Experimental Neurology, 61*(1), 1–11.

Rosen, G. D., Bai, J., Wang, Y., Fiondella, C. G., Threlkeld, S. W., LoTurco, J. J., & Galaburda, A. M. (2007). Disruption of neuronal migration by RNAi of Dyx1c1 results in neocortical and hippocampal malformations. *Cerebral Cortex, 17*(11), 2562–2572. doi:10.1093/cercor/bhl162

Rosen, G. D., Herman, A. E., & Galaburda, A. M. (1999). Sex differences in the effects of early neocortical injury on neuronal size distribution of the medial geniculate nucleus in the rat are mediated by perinatal gonadal steroids. *Cerebral Cortex, 9*(1), 27–34.

Rosen, G. D., Sherman, G. F., & Galaburda, A. M. (1991). Ontogenesis of neocortical asymmetry: A [3H]thymidine study. *Neuroscience, 41*(2-3), 779–790.

Rosen, G. D., Sherman, G. F., & Galaburda, A. M. (1992). Biological substrates of anatomic asymmetry. *Progress in Neurobiology, 39*(5), 507–515.

Szalkowski, C. E., Fiondella, C. G., Galaburda, A. M., Rosen, G. D., Loturco, J. J., & Fitch, R. H. (2012). Neocortical disruption and behavioral impairments in rats following in utero RNAi of candidate dyslexia risk gene Kiaa0319. *International Journal of Developmental Neuroscience, 30*(4), 293–302. doi:10.1016/j.ijdevneu.2012.01.009

Truong, D. T., Che, A., Rendall, A. R., Szalkowski, C. E., LoTurco, J. J., Galaburda, A. M., & Holly Fitch, R. (2014). Mutation of Dcdc2 in mice leads to impairments in auditory processing and memory ability. *Genes Brain and Behavior, 13*(8), 802–811. doi:10.1111/gbb.12170

Von Economo, C., & Horn, L. (1930). Über Windungsrelief, Masse, und Rindenarchitektonik der Supratemporalflache, ihre individuellen und ihre Seitenunterschiede. *Zeitschrift für die gesamte Neurologie und Psychiatrie, 130*, 678–757.

Wang, Y., Paramasivam, M., Thomas, A., Bai, J., Kaminen-Ahola, N., Kere, J.,…Loturco, J. J. (2006). DYX1C1 functions in neuronal migration in developing neocortex. *Neuroscience, 143*(2), 515–522. doi:10.1016/j.neuroscience.2006.08.022

Wang, Y., Yin, X., Rosen, G., Gabel, L., Guadiana, S. M., Sarkisian, M. R.,…Loturco, J. J. (2011). Dcdc2 knockout mice display exacerbated developmental disruptions following knockdown of doublecortin. *Neuroscience, 190*, 398–408. doi:10.1016/j.neuroscience.2011.06.010

CHAPTER 2

Clinical and Sociological Aspects of Dyslexia in the Context of the Geschwind-Galaburda Hypothesis 30 Years Later

David Urion

"The great tragedy of science—the slaying of a beautiful hypothesis by an ugly fact." Thomas Huxley (1870)

"A beautiful theory is better than an ugly fact." Norman Geschwind (personal communication, July 1, 1984)

SUMMARY

This volume is based on the history of clinical thinking about dyslexia and all the various field theories that preceded the work that was central to the hypothesis considered in the Extraordinary Brain conference. I used the technique of citation analysis and reviewed the clinical impact that a developmentally oriented, brain-based theory regarding the seminal human activity of reading had on the domain of dyslexia and, more generally, in pediatric behavioral neurology. I believe that the Geschwind-Galaburda Hypothesis (GGH) served as a point on which an entire area of child neurology and developmental studies pivoted, with important implications for developmental neuroscience since the mid-1980s and for the future. Research has come a long way, and the work continues to challenge and illuminate the field.

INTRODUCTION

A seminal series of papers appeared in the *Archives of Neurology* and made a bold series of speculations (Geschwind & Galaburda, 1985a, b, c). They offered a sort of field theory that related observations about a specific learning disorder—dyslexia—to mechanisms involved in establishing cerebral lateralization for a variety of cognitive tasks, including motor dominance, certain autoimmune disorders, and some phenomena involving ectodermal tissues. The papers also underlined male–female differences in the incidence and occurrence of these phenomena. That is, they posited a relationship among dyslexia, handedness, autoimmune disorders, and premature graying of the hair. The theory was based on a set of anatomic-pathological and clinical observations. The GGH, in its most basic form, states, "Intrauterine testosterone causes variations in lateralization (handedness and language), brain asymmetry, learning disabilities, learning superiority, immunological and other disorders, and male–female disorders" (A.M. Galaburda, personal communication, June 16, 2016). The papers proved to be hugely controversial—an explosion of writing followed that objected to their sweep and scope, some of the data cited, and the overall applicability in dyslexia research.

This volume considers that hypothesis 30 years later and is based on the proceedings of a conference devoted to this topic. The past three decades were a period of explosive growth in the techniques and capacities of cognitive neuroscience, as well as substantial expansion in education interventions. This chapter considers the hypothesis from the perspective of clinical medicine and the sociological context of medicine since the papers were originally published.

I am primarily a clinician and medical educator, serving as the program director of the child neurology and neurodevelopmental disabilities residency training programs at Boston Children's Hospital, as well as the director of the postresidency fellowship in behavioral neurology that the behavioral neurology unit has sponsored for the last 30 years. I would like to consider these seminal papers from the vantage point of someone with a reasonably large practice in pediatric behavioral neurology as well as one who serves as a training director in clinical neuroscience.

I was invited many years ago to the predecessor conference to this series of symposia on the Extraordinary Brain—From Neurons to Reading. For anyone who was there, or anyone who read the monograph that arose from it, it was indeed an extraordinary collection of speakers, not unlike the group gathered for this symposium; but in the mid-1980s, with the GGH Wars in full swing, it also had the feeling of something a bit edgy, almost clandestine, because of the controversy surrounding the GGH. It felt like the meeting of a persecuted sect at times, given the reactions more established workers in the field had to the hypothesis.

MY PATH TO NEUROLOGY

I came to pediatric behavioral neurology in a somewhat circuitous fashion. I was a French major in college and was writing my undergraduate thesis on pre-revolutionary French theater in its sociocultural context—a Gramscian Marxist analysis. I was housed with a family in Strasbourg, and the father was a behavioral neurologist who had studied under Jean-Louis Signoret at the Salpêtrière Hospital in Paris. He was translating the Boston Diagnostic Aphasia Test (Goodglass & Kaplan, 1972) from English into French and needed a colloquial North American English speaker to make sure the real sense of ideas had been captured. I signed on and went out to observe the field tests of the translated instrument.

Noticing how intrigued and engaged I seemed, he asked if I had ever considered going to medical school and becoming a neurologist. "Well, not really," I said. "You should think about it," he said. "You seem to take to it."

I went back, retread as pre-med, and then got into Stanford University, which was looking to have intellectually diverse backgrounds in its medical school classes. A French major qualified on that count. I arrived with the intent of becoming a behavioral neurologist for children because I had always had an affinity for working with a younger population.

Finding Norman Geschwind

I wanted to come to the Longwood Area Neurology Training Program at Harvard Medical School for essentially one reason—Norman Geschwind. I had read *Disconnexion Syndromes in Animals and Man* (Geschwind, 1965) and thought, "This is what I want to do, and this is with whom I want to study." I reasoned that, even though he was a neurologist for adults, his worldview and his papers made more sense to me than anything being written on the pediatric side. I thought that the application of his thinking could manifest in something applicable to children if I could be near him and learn from him. It was 1980.

The interview process was a bit daunting. I thought my interview at the Peter Bent Brigham Hospital did not go very well. I had yet to understand the enigmatic style of Charlie Barlow at Boston Children's Hospital, so I was a bit anxious as I walked over to Beth Israel Hospital. I was crestfallen when I was told that Dr. Geschwind was out of town. Simeon Locke interviewed me. If you never crossed his path, then let me explain that he raised the acerbic remark to an art form otherwise unknown in his generation of neurologists.

"Mr. Urion," he said, "I have just one question for you. Do you believe that higher cortical functions can be localized?" Was this a trick question? I was at the Beth Israel Hospital—the Vatican of cognitive and behavioral neurology. I smiled confidently. "Of course, Dr. Locke."

"Oh God," he said, "Another one."

In any event, I was somehow accepted and spent the next 5 years listening to Geschwind, Marsel Mesulam, and a very young and brilliant neurologist named Albert Galaburda.

The Hypothesis

I finished my training in 1985, and the GGH papers appeared in the Archives of Neurology during the course of that year. They addressed cerebral lateralization, biological mechanisms, associations, and pathology. I read them avidly, but in a way they were not really a surprise. I had been hearing the major tenets of the hypothesis since 1980. They were elegant, seamless, and synthesized ideas into a framework. I had come to accept these theories as the best way to describe how an important set of events in brain development occurred and how it was related to seemingly unusual coextant conditions. The theories explained why so many of my patients also saw immunologists or endocrinologists, why so many were left handed or ambidextrous, and why I could suss out the family in the waiting room at a first visit—it was typically the child who was seated with the nice 30-year-old father with a shock of completely white hair (a sign of a focal immunological attack on the pigment-making cells in the hair follicle).

I was mildly surprised when the controversy and pushback first occurred. It seemed a bit like arguing about gravity after Newton. I became befuddled as the ferocity of the disagreements accelerated. Was this petty jealousy in a field in which no one had had such a bold idea for decades? Much of the work up to this point had been describing associations and phenomena but had no overarching biological principles. It seemed a bit like those scribes who were mistrustful of the printing press because it would put them out of work.

In any event, the hypothesis did not go away, and the nastier bits of the furor eventually began to recede. As John Adams pointed out, "Facts are stubborn things; and whatever may be our wishes, our inclinations, or the dictates of our passion, they cannot alter the state of facts and evidence" (1770). And hypotheses that can be tested, challenged, examined, and evolve based on those sorts of interactions do tend to stay around.

Much of the rest of this volume will consider how the GGH has weathered the test of time and will explore contemporary findings that have sprung from this fertile ground. I will consider the papers in their historical context as well and their impact on clinical practice.

THE HISTORICAL CONTEXT

Approximately 11,800 scholarly papers in peer-reviewed journals were written on dyslexia from 1955 to 1985. That number rose to 78,300—a nearly seven-fold increase—from 1985 to 2015. Approximately 2,260 of the papers on dyslexia during the earlier period considered abnormalities or findings in the cerebral cortex in some fashion; that number jumped to 22,800—a 10-fold increase—after 1985.

This could be the evolution of the neurosciences in general, or behavioral neurology in particular, but let us compare another disorder—temporolimbic epilepsy (TLE). The same scholarly journals had 78,700 articles on TLE from 1955 to 1985, based on a MedLine search. That number was 82,000 from 1985 to 2015, a miniscule increase. And if we consider those which addressed some anatomic aspect of cerebral cortical abnormalities or migration findings, the number that addressed some anatomic aspect of cerebral cortical abnormalities was 10,900 from 1955 to 1985 and 26,800 from 1985 to 2015, a 2.4-fold increase (see Figure 2.1).

I obviously cannot prove that the Geschwind and Galaburda (1985a, b, c) papers were a cause, let alone a major cause, of this shift in publication pattern. But it is remarkable that 1985 was an apparent inflection point in this literature in which there was a shift away from papers written about dyslexic phenomena, examination findings, or description of clinical

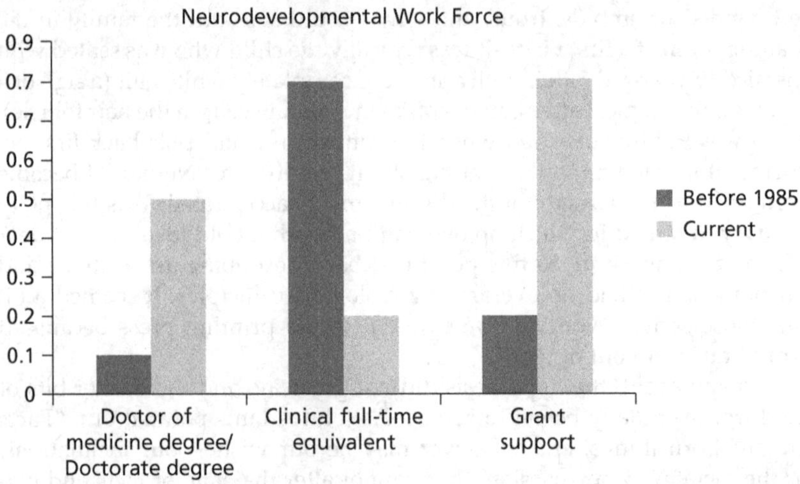

Figure 2.1. Fraction of work force for various characteristics, comparing periods before 1985 and current. Fraction that held a doctor of medicine degree and a doctorate degree, mean clinical full-time equivalent, and fraction that had any external grant support for their base salary.

elements and a shift toward a deeper dive into the anatomy and biology of the disorder. The field moved from a syndromic definition of the disorder (a person has dyslexia because of clinical findings) toward a biologic diagnosis (a person with dyslexia likely has certain biologic features present).

Changes in the Child Neurology Work Force The nature of those attracted to child neurology is a second aspect that should be considered in the clinical context of the hypothesis. What sort of people does the clinical domain of pediatric behavioral neurology attract now as compared with the past? I am currently serving as the president of the Professors of Child Neurology (PCN), a somewhat musty name for the organization within North American Child Neurology that is directly concerned with the care and well-being of academic programs in child neurology, and, in particular, the training programs that create the next generation of child neurologists. The Child Neurology Society (the more general organization of all child neurologists) and the American Academy of Pediatrics conduct a census of the profession approximately every decade. How many child neurologists are there? How are they deployed? What are they doing? How happy are they in their careers? The most recent survey was completed in 2015. We can use this data, which was extracted from a larger survey for the PCN, to look back over the last 40 years and examine trends specific to academic child neurologists.

It has been well established that the field of child neurology attracted an increasing number of physician-scientists to join its ranks starting in the late 1980s and early 1990s. This is reflected in the increased percentage of child neurologists who hold a doctor of medicine degree and a doctoral degree (fewer than 10% in the period before 1985; currently approximately 35%) as well as those who characterize their work as having at least 40% effort engaged in some form of research (up from 5% in the period before 1985 to more than one third in the recent survey). This means that child neurology has become an increasingly academic-trophic enterprise, with academic medical centers dominating that particular landscape. We can see this quite clearly in my home state of Massachusetts. There are two child neurologists left in private practice in the Commonwealth, and both are older than I. The entire rest of the child neurology work force has an academic affiliation. Another way to examine this is to compare the clinical time of child neurologists retiring from the work force with those entering. The average clinical full-time equivalent (FTE), which is the portion of one's work week spent in an activity, is 0.7; those entering average 0.4 (Kang et al., 2016).

Although this has many implications for the health of America—a topic for another time—I offer background information so we can consider what has specifically happened to the pediatric behavioral neurology work force over a similar period. Although those numbers are quite small compared with larger subspecialties such as epilepsy (10% of the work force self-identifies as having a practice that has a large component of neurobehavioral disorders compared with nearly one third that make a similar claim to epilepsy), it is perhaps illuminating to examine who those people are and what they do. More than half of those involved in neurodevelopmental disorders and behavioral neurology have a clinical doctor of medicine degree and a research doctoral degree, their average clinical FTE is 0.2, and 80% are engaged in grant-supported research (see Figure 2.2).

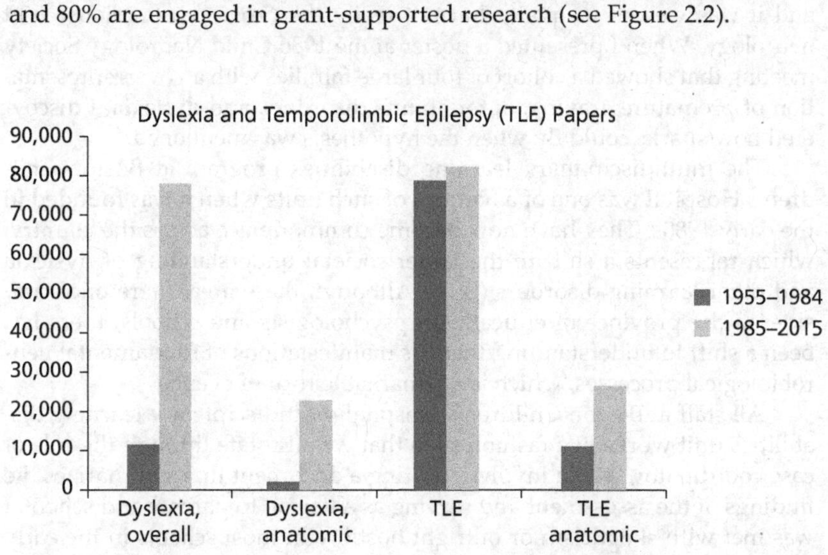

Figure 2.2. Number of papers in peer-reviewed journals in the periods 1955–1984 versus 1985–2015.

The work force in this domain is very scientifically engaged and driven. Young neuroscientists are drawn to a field that is now seen as fertile ground for their investigation. If we segregate out the behavioral neurology/neurodevelopmental disorders part of the child neurology work force and examine how most answer the question, "What attracted you to your area of subspecialty?" close to 80% state that the draw was scientific curiosity and the excitement of likely discoveries.

Compare this with similar data from surveys before 1985, in which 10% held a clinical doctor of medicine degree and a research doctoral degree, the average clinical FTE was 0.8, and fewer than 20% were in grant-supported research. Many were drawn to the field because of a personal experience (e.g., a close relative with a similar disorder) (Child Neurology Society, 1985).

Again, this reflects that the field is viewed as an exciting and attractive one for young clinician-scientists, which it was not before the GGH. I cannot say that the three papers we are here to reflect upon were a cause, let alone a major cause, of this shift, but those papers were emblematic of a seismic shift in the nature of the field and the work force.

Changes in Practice Finally, I would like to consider how these papers changed the way those of us in practice do just that—practice. My training program in 1982 was one of the few that taught child neurologists to ask questions about literacy, reading acquisition, and school performance. As a young child neurologist in training I would bring posters to meetings about ideas such as automaticity being afflicted in both stuttering and dyslexia, and it was viewed as quaint but not really the province of serious child neurology. When I presented a poster at the 1986 Child Neurology Society meeting that showed a cohort of four large families with an overrepresentation of premature graying, autoimmune disorders, and dyslexia, I discovered how sparks could fly when the hypothesis was mentioned.

The multidisciplinary learning disabilities program at Boston Children's Hospital was one of a handful of such units when it was founded in the early 1980s. They have now become commonplace across the country, which represents a shift in the larger societal understanding of dyslexia and other learning disorders (LDs). Although these areas were once considered the province of educational psychologists and schools, there has been a shift to understanding them as manifestations of fundamental neurobiological processes, which is a remarkable turn of events.

All staff at Boston Children's Hospital multidisciplinary learning disabilities unit works in a fashion such that we all rotate through the role of case coordinator, which involves writing a document that summarizes the findings of the assessment and serving as a liaison to parents and school. I was met with skepticism or outright hostility in most schools in the early years of my career. "What," I was often archly asked, "does a physician know about reading?"

The times have changed. Not very long ago, the head of the reading program at the University of New Hampshire School of Education was on a panel discussing diagnosis, definition, and implications for policy. He said, "We can identify a reading disorder. But dyslexia is a medical diagnosis and requires a medical assessment." Pendulums sometimes swing a bit far but reflect current understanding in an academic cultural context. One of the effects of the papers, the hypothesis, and the subsequent work has been the medicalization of the diagnosis of dyslexia, with all that that implies.

We can view board examinations for specialty certification and the tools that prepare residents to take those examinations as a reflection of what the profession feels is the corpus of knowledge necessary for the practice of the specialty. The Residency In-Service Training Examination (RITE) is a self-assessment tool the American Academy of Neurology produces to help trainees assess their progress in the body of knowledge needed to practice neurology, which, in turn, is reflected by the written examination for board certification. Questions regarding reading and LDs began to appear on the RITE in the mid-2000s; although still small in number, the very fact that they appeared reflects an acknowledgment of the relevance of the area. Because the exam is excruciatingly time-choreographed (every question must earn its place in the pantheon, based on reliability and relevance), this is further testimony to the field being mainstream. Again, I cannot prove that the GGH is related to this, but it surely contributed to the gravitas of the area and, hence, the inclusion in an instrument which by its nature is conservative (i.e., preparation for a board examination).

Residency training is time-pressured by hours limitations and the multiple demands placed on clinical teachers. They must see increasing numbers of patients themselves, get residents out in a timely fashion, and fill out mountains of electronic forms that allege meaningful use. I think back on my residency days, when we would all come in for Saturday morning rounds at Beth Israel Hospital. The real day began after the patients were seen and tucked in. A pot of coffee was appropriated from the nurses' station, and Norman would sit down with us and start to talk. Lively, engaged, and often joined by brilliant younger faculty, he would regale us with stories, pose questions, and hold forth on all things in behavioral neurology, and we would hear many of the salient features of the hypothesis. The sacred texts were discussed, analyzed, argued over, and reasoned with. A very Talmudic experience.

Gone are those days. In order to help our colleagues try to maintain some semblance of a comprehensive educational program, the Consortium of Neurology Program Directors, which includes both adult and child neurology program directors, maintains a web-based tool kit from which teaching materials can be downloaded for use in a program. A large number of modules have been created, and the American Academy of Neurology, which is a data-driven organization, counts the number of hits each module gets. The Evaluation of Reading and Language Disorders is a module to which our program contributed and is fundamentally the embodiment of the assessment for clinical elements associated with the main tenets of the hypothesis, as well as office-based assessment tools and a videographed lecture on the assessment with examples of each tool in use. That module is routinely the third most downloaded from the web site and the most downloaded of all the pediatric modules. It is only surpassed by modules on acute stroke management and the management of status epilepticus. Again, I cannot prove that this is due to the papers discussed in this volume (Geschwind & Galaburda, 1985a, b, c), but it is remarkable that this area of interest has become so mainstream that time-starved teaching programs value it so manifestly.

I fundamentally understand my work as a clinician through stories. Diagnosis comes from the Greek *dia*—through *gnosein*—"to know." One knows things through stories. Our residency program found itself with some unanticipated openings. Child neurology matches (meaning candidates match to programs for their residencies) 2 years in advance, and one year several people found that they wanted to stay as general pediatricians and not subspecialize. We found ourselves with openings and filled the empty spots in our residency training program with people who had finished 3 years of general pediatric training—thus, field-tested, combat-hardened, pratical people.

One of the residents had trained at another large children's hospital in an urban setting, a busy place that serves a hard-put urban population alongside exquisite tertiary and quaternary care. This man was tough, competent, and pragmatic. He was presenting a patient to me in general neurology clinic. He started, "So this little 5-year-old boy is left handed by Oldfield criteria..." I stopped him, as is a teacher's prerogative. "Why do we care about handedness?" He smiled. He was getting used to the more expansive discussions our hospital permits without the press of 17 patients waiting. "Well," he said, "if there is no family history of left-handedness, because it can be a genetic trait in part, then it makes us alert to the possibility of early injury and transfer of motor dominance to the right hemisphere. But if there is a strong family history of left-handedness, we then need to look for other factors, such as immune disorders and learning problems. And it increases the child's risk for those things, so we have to follow them closely. Oh, and sometimes people in the family go gray early. We need to ask about that. No disrespect, DKU." (Most of our residents refer to me by my initials. I generally discourage them from calling me Dr. Urion and offer 'David'; DKU is a compromise.)

It was my turn to smile. "How do we know that?"

He looked puzzled. "You mean, why are those things true? I don't know. It's like a Babinski. It's just true."

In case you are not a clinician, Babinski's sign is the fanning of the toes and the upward movement of the great toe when the sole of the foot is stroked, which denotes corticospinal tract injury on that side. It is the only truly incontrovertible physical sign of the neurologic examination first described by one of Jean Martin Charcot's ablest trainees, Joseph Jules Francois Felix Babinski. Babinski was a teacher to Paul Yakovlev, famous for his studies on myelogenesis of the developing human brain, who, in turn, taught Norman Geschwind, who taught Galaburda, who taught me.

This young man, a product of an excellent but very practical pediatric training program, whose own goal is to become a pediatric epileptologist and return to his region of origin to improve health care there, who had no knowledge of my own path through neurology, and who had relatively little sense of the history of our training program and its ancient association with the Beth Israel of Geschwind and Galaburda, who does not have an ounce of "gee Professor tell us that story again" lickspittle about him, if this young man accepts that the major tenets of the GGH are as central to his chosen profession as Babinski's sign, then we have come a long way since the mid-1980s.

This is dharma, not dogma. *Dogma* is usually defined as a belief accepted without question or doubt. *Dharma* is a term used by a number of traditions, but in its Buddhist manifestation it refers to the teachings of the Buddha, in which one was always encouraged by the Buddha to take the teachings and measure them against one's own experience. Dharma comes from the Sanskrit *dhr*—to hold, to maintain, to preserve, as for future use. This volume contains some of the dharma—how a variety of neuroscientists, practitioners, and educators have all taken the received theory, investigated it, challenged it, extended it, and confronted it. They have tested the tensile strength it adds to the fabric of contemporary neuroscience and, I would submit, found that to be considerable.

REFERENCES

Adams, J. (1770). *Argument in defense of the soldiers in the Boston Massacre trials, December 1770.* Retrieved from http://www.quotationspage.com/quote/3235.html

Child Neurology Society. (1985). *CNS/AAP survey data.* Retrieved from http://www.childneurologysociety.org/pcn/archive

Geschwind, N. (1965). *Disconnexion syndromes in animals and man.* Retrieved from http://dx.doi.org/10.1093/brain/88.2.237

Geschwind, N., & Galaburda, A. M. (1985a). Cerebral lateralization. Biological mechanisms, associations, and pathology: I. A hypothesis and a program for research. *Archives of Neurology, 42*(5), 428–459.

Geschwind, N., & Galaburda, A. M. (1985b). Cerebral lateralization. Biological mechanisms, associations, and pathology: II. A hypothesis and a program for research. *Archives of Neurology, 42*(6), 521–552.

Geschwind, N., & Galaburda, A. M. (1985c). Cerebral lateralization. Biological mechanisms, associations, and pathology: III. A hypothesis and a program for research. *Archives of Neurology, 42*(7), 634–654.

Goodglass, H., & Kaplan, E. (1972). *The assessment of aphasia and related disorders.* Boston, MA: Lea & Febiger.

Huxley, T. (1870). *Biogenesis and abiogenesis.* Presidential address to the British Association for the Advancement of Science meeting in Liverpool. *Nature,* 2(400–406).

Kang, P. B., Bale, J. F., Mintz, M., Joshi, S. M., Gilbert, D. L., Radabaugh, C., & Ruch-Ross, H. (2016). The child neurology clinical workforce in 2015: Report of the AAP/CNS Joint Task Force. *Neurology, 87*(13), 1384–1392.

SECTION II

Brain Development, Hormones, and Immunology

CHAPTER 3

Mechanisms that Control Left/Right Asymmetry and Sexual Dimorphisms in the Forebrain

John L.R. Rubenstein

SUMMARY

The Geschwind-Galaburda Hypothesis (GGH) provided an important impetus to identify mechanisms that control brain left/right asymmetry and sexual dimorphisms, particularly as they relate to language and reading processing systems. Developmental, genetic, hormonal, and stochastic processes may contribute to the development of language processing circuits and could underlie some developmental disabilities, including dyslexia. It is important to consider the following when studying the mechanisms underlying language disorders, including dyslexia: 1) all circuit components (even peripheral ones) needed for language development, 2) long distance and local circuits, 3) alterations to synapse numbers and/or strengths in those circuits, and 4) the development of white matter (myelination) along those circuits.

INTRODUCTION

I was a resident in psychiatry at Stanford University in 1987, and I took a course from Marsel Mesulam and Albert Galaburda that may have been called The Neuroanatomy of Behavior. I was introduced to the GGH (Geschwind & Galaburda, 1985a, b, c) during the course, which was a formative experience for me. I learned about brain structure and function, particularly as it related to higher cognitive functions, including language. Learning about asymmetry of brain functions (lateralization) and brain anatomy (e.g., the planum temporale) led me to consider research projects that could help identify molecular and developmental mechanisms underlying the generation of these asymmetries. I strongly considered approaching this using subtractive hybridization to identify genes that are expressed at different levels in the left and right side of the telencephalon (developing cerebrum). Although I did not perform that experiment, I did perform a related one that laid the foundation for nearly 30 years of my subsequent research on the identification of genes that regulate regional and cellular identity in the developing telencephalon. Since the mid-1980s,

This work was supported by the research grants to JLRR from Nina Ireland and NIMH R37 MH049428 from the National Institutes of Health.

I have gained perspective about approaches to studying problems in developmental neurobiology and have maintained an active interest in human neurodevelopmental disorders, such as mental deficiency, autism spectrum disorder, schizophrenia, and dyslexia. Thus, I wanted to contribute as a participant in the Extraordinary Brain Series Symposium XV: The Geschwind-Galaburda Hypothesis 30 Years Later by providing some suggestions for possible developmental, neurobiological, and genetic mechanisms that may underlie some forms of dyslexia.

DIFFERENCES IN THE BRAINS OF MALES AND FEMALES

A postulate of the GGH is that there are male–female differences in brain structure and function, particularly in the parts of the neocortex involved in language. These differences may underlie the increased prevalence of cognitive and behavioral disorders observed in boys over girls, including dyslexia. Furthermore, it is postulated that sex hormones, testosterone in particular, may have a fundamental role in creating this bias.

We learned of male–female differences in several brain features during the June 2016 meeting, including sex differences in cerebellar gray matter asymmetry in lobules VII, VIII, and IX. Geert de Vries described multiple sex differences in vasopressin innervation within the brain that contribute to differences in physiology and behavior (De Vries & Forger, 2015; see Chapter 4). Furthermore, abnormal numbers of sex chromosomes (e.g., X0, XYY) show changes in the relative sizes of brain regions, including the amount of neocortical gray matter (Lepage et al., 2013). Although the mechanisms for these human neuroanatomical differences are not known, experiments in rodents demonstrated that differential neuronal survival accounts for male–female differences in the size of particular structures, including the median preoptic nucleus (Ngun, Ghahramani, Sánchez, Bocklandt, & Vilain, 2011). Hypothalamic nuclei, such as the ventromedial hypothalamus (VMH), have multiple subdivisions. A subset of VMH neurons in the ventrolateral part (VMHVL) express estrogen receptor-alpha; the VMHVL regulates specific behaviors, such as estrogen-dependent locomotion. Disrupting the development of estrogen receptor-alpha neurons in the VMHVL, by deletion of the Nkx2-1 transcription factor gene, results in female-specific obesity (Correa et al., 2015). Other sexually dimorphic neurons in the VMH regulate aggression in males (Yang et al., 2013). Thus, many examples of sexually dimorphic subdivisions in the rodent forebrain govern behavioral differences between males and females. It seems likely that the human brain has similar sexually dimorphic brain regions, which could underlie some differences in male and female behaviors.

ORIGINS OF LEFT/RIGHT ASYMMETRIES: A CONSIDERATION OF HARD-WIRED AND STOCHASTIC MECHANISMS

Another postulate of the GGH is that the developmental origins of cerebral asymmetries contribute to lateralization of language functions; disruption of these developmental processes may contribute to cognitive problems of language. I first want to comment on developmental mechanisms that generate left/right asymmetry in vertebrates.

The origins of left/right body asymmetry are generated during gastrulation and neural plate stage in vertebrate embryos by leftward beating of cilia from cells in a midline structure called *the node*. Evidence suggests that asymmetric beating of cilia generates a flow of molecules that may be sensed by other cilia to generate left/right gene expression differences that create asymmetric development of the viscera (e.g., heart, lung, abdominal organs) (Blum, Feistel, Thumberger, & Schweickert, 2014). Analyses of mutant mice and fish that demonstrate left/right asymmetries of body organs have identified genes that regulate this process, including the nodal secreted protein, the lefty and pitx2 transcription factors, and the Pkd channels.

The epiphysis and habenula, which form in the dorsal diencephalon are another set of structures that develop a left/right asymmetry in vertebrate embryos. The habenula is part of pathways that interconnect the limbic forebrain with loci in the midbrain and hindbrain that control dopaminergic and serotonergic pathways (Bianco & Wilson, 2009). The left and right dorsal habenula in fish have asymmetric patterns of gene expression and asymmetric efferent and afferent projections. Evidence shows this asymmetry contributes to the functions of these structures, such as responses to visual and olfactory stimuli (Dreosti, Vendrell Llopis, Carl, Yaksi, & Wilson, 2014). Habenular function is linked to many other processes, including the control of dopaminergic modulation of motor activity and reward prediction, spatial learning and attention, responses to aversive stimuli, and human neuropsychological disorders (Bianco & Wilson, 2009). As in the node, signaling by the nodal protein is implicated in the generation of habenular left/right asymmetry (Concha, Bianco, & Wilson, 2012). Function of an autism risk gene, the Tcf7l2 transcription factor, has been show to regulate habenular left/right asymmetry, providing evidence that Wnt-signaling is also involved in this process (Hüsken et al., 2014).

Efforts to decipher mechanisms underlying left/right asymmetries in the human cerebral cortex, using differential gene expression methods in fetal brains, have yielded a number of candidate genes. One named Lmo4 shows the most consistency (Sun & Walsh, 2006). Lmo4 is a transcription factor that is required in mice for controlling the identity of layer five motor cortex projection neurons (Cederquist, Azim, Shnider, Padmanabhan, & Macklis, 2013). Studies of left/right asymmetries in the rodent brain are hampered because left/right behavioral asymmetries (e.g., paw preference) are not observed at the population level, but they do exist in individual animals (Sun & Walsh, 2006). Thus, left/right asymmetry in the rodent cortex may be generated randomly (Sun & Walsh, 2006). In this regard, it is important that loss of nodal signaling at the node leads to randomization of left/right asymmetry of body organs. Thus, creating the bias in human left/right brain asymmetry may involve an evolutionary gain of a molecular mechanism.

The origin of asymmetry of cortical functions in rodents, albeit lacking a systematic, population-level, left/right bias, is unknown. Perhaps embryonic patterning centers that govern early steps in the generation of regional identity could contribute to left/right asymmetry in humans. The rostral patterning center secretes proteins of the FGF family, including FGF8 and FGF17, in the embryonic neural plate and rostral telencephalon (Cholfin & Rubenstein, 2007; Shimamura & Rubenstein, 1997). These proteins specify rostral identity to the cerebral cortex and basal ganglia (Cholfin & Rubenstein, 2007; Fukuchi-Shimogori & Grove, 2001; Garel, Huffman, & Rubenstein, 2003; Huffman, Garel, & Rubenstein, 2004; Storm et al., 2006). Although the pattern of FGF8 expression in the rostral patterning center has not shown asymmetry, fate mapping of cells that express FGF8 at the neurulation stage in mice show left/right asymmetries in their numbers in the cerebral cortex (see Figure S3 in Hoch, Clarke, & Rubenstein, 2015). Although the functional significance for this preliminary observation has not been established, it could shed light on basic developmental mechanisms underlying left/right asymmetries in the neocortex.

STOCHASTIC PROCESSES IN BRAIN DEVELOPMENT

If left/right cortical asymmetry were governed by stochastic mechanisms in rodents, then it is possible that the magnitude of the anatomical asymmetry, or the effect of the asymmetry, could be also regulated by a stochastic process that is under the control of distinct molecular mechanisms. Although most of brain development seems to follow rigid genetic rules that are essential to create species-specific nervous systems, stochastic processes may be layered on top of this to generate structural and functional diversity that leads to variations in cognition and behavior with the possibility of positive adaptive properties. Of course, the converse of this would be the generation of negative adaptive properties, perhaps contributing to brain disorders. Thus, on a population level, individuals at the extremes of the distribution of brain functions (of various domains) may, in part, be the result of stochastic process in brain development and/or function. Molecular and/or environmental variables that increase the probability and/or magnitude of the effect of the stochastic process could, thereby, contribute to generating individuals with either deficits or gains in particular brain functions.

Stochastic processes could affect any developmental and/or brain function processes. In addition to the stochastic left/right cortical fate mapping result (Hoch et al., 2015), documented random processes take place during brain development. Important advances show that mutations in our genome take place in germ cells, stem cells in the embryo (during DNA replication), and somatic cells (perhaps taking place during transcription), which can lead to the accumulation of thousands of single nucleotide variants (SNVs) in addition to larger mutations (e.g., copy number variants;

CNVs) in cells of our brains (Lodato et al., 2015). Mutations are also caused by retrotransposition events by Line Element 1 (L1) (Evrony et al., 2015; Muotri et al., 2005; Upton et al., 2015). For instance, Evrony et al. reported a mutation frequency of approximately 0.2 events per brain cell tested using single cell analysis methods. Furthermore, there is evidence for increased L1 transposition (e.g., induced mutations) in cells lacking the Rett syndrome gene *(MeCP2)* (Muotri et al., 2010), which can be a mechanism by which defects in epigenome regulators alter the genome and brain function.

Mutations that take place as the brain develops can be detrimental. For instance, a somatic mutation in developing brain cells causes a gain of function in the product of the AKT3 kinase gene leading to hemimegalencephaly (HMG; unilateral enlargement of cerebral hemisphere associated with epilepsy) (Poduri et al., 2012). HMG is related to another set of disorders called *focal cortical dysplasias* (FCD; small abnormal local growths within the brain); these are features in individuals with mutations in the TSC1&2 genes that cause tuberous sclerosis (D'Gama et al., 2015). AKT3 and TSC1&2 generate proteins that regulate signaling in a critical biochemical pathway that controls cell growth, known as the mTOR pathway. It is now appreciated that mutations in additional genes in this pathway cause HMG and focal cortical dysplasia (D'Gama et al., 2015). Therefore, genetic, chemical, and/or environmental factors that increase mutation rates during development are important factors that increase the probability for variance in genome and brain function and individual differences, even between identical twins.

GENETICS AND DEVELOPMENT

Up to this point, I have largely focused on processes that control regional identity, proliferation, and cell survival during development. Many other processes are required to generate the central nervous system, whose derailment could lead to brain dysfunction that could contribute to language and communication disorders. I do not want to address all of these in detail, but it is worth considering that defects in cell type specification (e.g., making excitatory and inhibitory neurons, astrocytes, and oligodendrocytes), cell migration, neural maturation of processes (e.g., dendrite, axons) and electrophysiological properties, axon pathfinding, synapse formation, and myelination are all required to generate brain circuits (local and distributed) that can interpret and execute the complex computations for the comprehension and generation of speech and language. I hesitate to posit how many genes could be involved in these processes, but identifying genes that increase the risk for dyslexia is the key entry point for understanding this disorder.

Progress has been made in defining the genes that contribute to dyslexia risk, including *CYP19A1, DCDC2, DIP2A, DYX1C1, GCFC2* (also known as *C2orf3*), *KIAA0319, MRPL19, PCNT, PRMT2, ROBO1,* and *S100B* (Matsson et al., 2015). Although additional studies are needed to firmly establish that these genes confer risk, *ROBO1* is known to regulate cell proliferation, axon pathfinding, and cell migration (López-Bendito et al., 2007). *DCDC2, DYX1C1,* and *KIAA0319* are implicated in cilia function (Ivliev, 't Hoen, van Roon-Mom, Peters, & Sergeeva, 2012). As previously mentioned, cilia are required in the node to generate left/right asymmetry during gastrulation. Dyslexia genetic studies may be underpowered, however, with respect to dyslexia's genetic heterogeneity/complexity. Future dyslexia genetic studies could become as successful as those in autism by using study designs that can identify de novo loss of function mutations (Geschwind & State, 2015) or study designs with sufficient individuals who have identified single nucleotide polymorphisms (SNPs) associated with schizophrenia (Schizophrenia Working Group of the Psychiatric Genomics Consortium, 2014).

One of the biggest challenges in understanding the genetics of complex human disorders is that mutations in noncoding regions of genes could contribute much of the risk. Although there are roughly 20,000 genes in the human genome, each gene is regulated by multiple regulatory elements, most of which are thought to function as enhancers. Enhancers are usually relatively small DNA sequences (~500–5000 base pairs) that can be within the gene (in introns), close to the gene, or far from the gene (1,000,000 base pairs away) (Nord, Pattabiraman, Visel, & Rubenstein, 2015). They can even be in another gene. Enhancers integrate the past and present history of the cells through epigenetic signatures on the DNA and on their chromatin, the combination of transcription factors present in the cell, and the effect of biochemical signals affecting the cell (Nord et al., 2015). For instance, enhancers are active in specific regions of the embryonic forebrain (Visel et al., 2013), including small subdivisions of the cerebral cortex such as the medial prefrontal cortex (Pattabiraman et al., 2014). These enhancers are regulated by transcription factors, such as PAX6, that control regional patterning of the cerebral cortex (Pattabiraman et al., 2014); the expression of these transcription factors is controlled by the patterning centers (e.g., the rostral patterning center that expresses FGF genes), which were previously discussed. Thus, cortical development is controlled by a genetic/developmental program that links patterning centers, transcription factor expression levels, and enhancers. I point this out to illustrate the complexity of brain development. Yet, once a gene or enhancer is implicated in a disorder, the field can gain traction in understanding the disorder if the function of the gene or enhancer is known in the context of brain development and/or brain function.

FOUR ADDITIONAL MECHANISMS TO CONSIDER

I would like to end by briefly addressing four processes that regulate brain development and function that could affect dyslexia. First, although it seems most likely that defects in the neocortex underlie the disorder, one needs to consider all components of the circuits that are required for speech and language. For instance, asymmetry of language processing, as well as defects in language decoding and comprehension, could arise at the earliest parts of the circuits, including in the inner and middle ear (cochlea and ossicles). Subtle differences in the size and/or function of this structure could underlie asymmetries and/or processing defects that could be propagated along the pathway.

Second, in addition to connectivity defects in these circuits, one needs to consider defects in the local circuit dysfunction. For instance, if there is a deficiency of inhibitory interneuron function, then this could initially create excessive local excitation, which may be followed by compensatory mechanisms that decrease excitatory response; both changes in the local circuitry could be detrimental (Nelson & Valakh, 2015; Rubenstein & Merzenich, 2003). Indeed, reduced numbers of inhibitory interneurons in the visual and auditory cortex alter their functions (Mao et al., 2012; Seybold et al., 2012).

Third, alterations to synapse numbers or strengths could affect both long distance and local circuits. In that regard, evidence shows that a variant of the NMDA glutamate receptor 2b is associated with language lateralization (Ocklenburg et al., 2011). Furthermore, extensive literature provides evidence that sex steroid hormones (testosterone and estrogen) can modify synapse and circuit functions (Lee & Pfaff, 2008; McEwen & Alves, 1999; Sellers, Raval, & Srivastava, 2015), controlling features such as the numbers of synaptic spines (Garelick & Swann, 2014) and mossy fiber transmission and LTP in hippocampal area CA3 (Skucas et al., 2013).

A final mechanism relates to the molecular and environmental processes that control the relative developmental timing and magnitude of myelination of specific pathways. Myelination potently increases action potential speed along axons and, therefore, ultimately strengthens the signaling and timing between various circuit components. Therefore, alterations in the degree of myelination could improve or degrade circuit functions. Given that MRI studies can assess the amount and directional coherence of white matter in specific pathways, and white matter reflects the degree of myelination, this dimension of brain structure, and possibly function, should be a reasonable window to grossly assess the status of brain circuits.

REFERENCES

Bianco, I. H., & Wilson, S. W. (2009). The habenular nuclei: A conserved asymmetric relay station in the vertebrate brain. *Philosophical Transactions of the Royal Society B Biological Sciences, 364,* 1005–1020.

Blum, M., Feistel, K., Thumberger, T., & Schweickert, A. (2014). The evolution and conservation of left-right patterning mechanisms. *Development, 141,* 1603–1613.

Cederquist, G. Y., Azim, E., Shnider, S. J., Padmanabhan, H., & Macklis, J. D. (2013). Lmo4 establishes rostral motor cortex projection neuron subtype diversity. *Journal of Neuroscience, 33,* 6321–6332.

Cholfin, J. A., & Rubenstein, J. L. (2007). Patterning of frontal cortex subdivisions by Fgf17. *Proceedings of the National Academy of Sciences USA, 104,* 7652–7657.

Concha, M. L., Bianco, I. H., & Wilson, S. W. (2012). Encoding asymmetry within neural circuits. *Nature Reviews Neuroscience, 13,* 832–843.

Correa, S. M., Newstrom, D. W., Warne, J. P., Flandin, P., Cheung, C. C., Lin-Moore, A. T.,...Ingraham, H. A. (2015). *Cell Reports, 10,* 62–74.

D'Gama, A. M., Geng, Y., Couto, J. A., Martin, B., Boyle, E. A., LaCoursiere, C. M.,... Poduri, A. (2015). Mammalian target of rapamycin pathway mutations. *Annals of Neurology, 77,* 720–725.

De Vries, G. J., & Forger, N. G. (2015). Sex differences in the brain: A whole body perspective. *Biology of Sex Differences, 6,* 15.

Dreosti, E., Vendrell Llopis, N., Carl, M., Yaksi, E., & Wilson, S. W. (2014). Left-right asymmetry is required for the habenulae to respond to both visual and olfactory stimuli. *Current Biology, 24,* 440–445.

Evrony, G. D., Lee, E., Mehta, B. K., Benjamini, Y., Johnson, R. M., Cai, X.,...Walsh, C. A. (2015). Cell lineage analysis in human brain using endogenous retroelements. *Neuron, 85,* 49–59.

Fukuchi-Shimogori, T., & Grove, E. A. (2001). Neocortex patterning by the secreted signaling molecule FGF8. *Science, 294,* 1071–1074.

Garel, S., Huffman, K. J., & Rubenstein, J. L. (2003). Molecular regionalization of the neocortex is disrupted in Fgf8 hypomorphic mutants. *Development, 130,* 1903–1914.

Garelick, T., & Swann, J. (2014). Testosterone regulates the density of dendritic spines in the male preoptic area. *Hormones and Behavior, 65,* 249–253.

Geschwind, D. H., & State, M. W. (2015). Gene hunting in autism spectrum disorder: On the path to precision medicine. *The Lancet Neurology, 14,* 1109–1120.

Geschwind, N., & Galaburda, A. M. (1985a). Cerebral lateralization. Biological mechanisms, associations, and pathology: I. A hypothesis and a program for research. *Archives of Neurology, 42*(5), 428–459.

Geschwind, N., & Galaburda, A. M. (1985b). Cerebral lateralization. Biological mechanisms, associations, and pathology: II. A hypothesis and a program for research. *Archives of Neurology, 42*(6), 521–552.

Geschwind, N., & Galaburda, A. M. (1985c). Cerebral lateralization. Biological mechanisms, associations, and pathology: III. A hypothesis and a program for research. *Archives of Neurology, 42*(7), 634–654.

Hoch, R. V., Clarke J. A., & Rubenstein, J. L. (2015). Fgf signaling controls the telencephalic distribution of Fgf-expressing progenitors generated in the rostral patterning center. *Neural Development, 10,* 8.

Huffman, K. J., Garel, S., & Rubenstein, J. L. (2004). Fgf8 regulates the development of intra-neocortical projections. *Journal of Neuroscience, 24,* 8917–8923.

Hüsken, U., Stickney, H. L., Gestri, G., Bianco, I. H., Faro, A., Young, R. M.,...Carl, M. (2014). Tcf7l2 is required for left-right asymmetric differentiation of habenular neurons. *Current Biology, 24,* 2217–2227.

Isaac, H., Bianco, I. H., Stephen, W., & Wilson, S. W. (2009). The habenular nuclei: A conserved asymmetric relay station in the vertebrate brain. *Philosophical Transactions of the Royal Society of London. Series B: Biological Sciences, 364,* 1005–1020.

Ivliev, A. E., 't Hoen, P. A., van Roon-Mom, W. M., Peters, D. J., & Sergeeva, M. G. (2012). Exploring the transcriptome of ciliated cells using in silico dissection of human tissues. *PLoS One, 7,* e35618.

Lee, A. W., & Pfaff, D. W. (2008). Hormone effects on specific and global brain functions. *Journal of Physiological Sciences, 58,* 213–220.

Lepage, J. F., Hong, D. S., Mazaika, P. K., Raman, M., Sheau, K., Marzelli, M. J.,...Reiss, A. L. (2013). Genomic imprinting effects of the X chromosome on brain morphology. *Journal of Neuroscience, 33,* 8567–8574.

Lodato, M. A., Woodworth, M. B., Lee, S., Evrony, G. D., Mehta, B. K., Karger, A.,...Walsh, C. A. (2015). Somatic mutation in single human neurons tracks developmental and transcriptional history. *Science, 350,* 94–98.

López-Bendito, G., Flames, N., Ma, L., Fouquet, C., Di Meglio, T., Chedotal, A.,...Marín, O. (2007). Robo1 and Robo2 cooperate to control the guidance of major axonal tracts in the mammalian forebrain. *Journal of Neuroscience, 27,* 3395–3407.

Mao, R., Schummers, J., Knoblich, U., Lacey, C. J., Van Wart, A., Cobos, I.,…Sur M. (2012). Influence of a subtype of inhibitory interneuron on stimulus-specific responses in visual cortex. *Cerebral Cortex, 22,* 493–508.

Matsson, H., Huss, M., Persson, H., Einarsdottir, E., Tiraboschi, E., Nopola-Hemmi, J.,…Kere, J. (2015). Polymorphisms in DCDC2 and S100B associate with developmental dyslexia. *Journal of Human Genetics, 60,* 399–401.

McEwen, B. S., & Alves, S. E. (1999). Estrogen actions in the central nervous system. *Endocrine Reviews, 20,* 279–307.

Muotri, A. R., Chu, V. T., Marchetto, M. C. N., Deng, W., Moran, J. V., & Gage, F. H. (2005). Somatic mosaicism in neuronal precursor cells mediated by L1 retrotransposition. *Nature, 435,* 903–910.

Muotri, A. R., Marchetto, M. C. N., Coufal, N. G., Oefner, R., Yeo, G., Nakashima, K., & Gage, F. H. (2010). L1 retrotransposition in neurons is modulated by MeCP2. *Nature, 468,* 443–446.

Nelson, S. B., & Valakh, V. (2015). Excitatory/inhibitory balance and circuit homeostasis in autism spectrum disorders. *Neuron, 87,* 684–698.

Ngun, T. C., Ghahramani, N., Sánchez, F. J., Bocklandt, S., & Vilain, E. (2011). The genetics of sex differences in brain and behavior. *Frontiers of Neuroendocrinology, 32,* 227–246.

Nord, A. S., Pattabiraman, K., Visel, A., & Rubenstein, J. L. (2015). Genomic perspectives of transcriptional regulation in forebrain development. *Neuron, 85,* 27–47.

Ocklenburg, S., Arning, L., Hahn, C., Gerding, W. M., Epplen, J. T., Güntürkün, O., & Beste, C. (2011). Variation in the NMDA receptor 2B subunit gene GRIN2B is associated with differential language lateralization. *Behavioural Brain Research, 225,* 284–289.

Pattabiraman, K., Golonzhka, O., Lindtner, S., Nord, A. S., Taher, L., Hoch, R.,…Rubenstein, J. L. R. (2014). Transcriptional regulation of enhancers active in protodomains of the developing cerebral cortex. *Neuron, 82,* 989–1003.

Poduri, A., Evrony, G. D., Cai, X., Elhosary, P. C., Beroukhim, R., Lehtinen, M. K.,…Walsh, C. A. (2012). Somatic activation of AKT3 causes hemispheric developmental brain malformations. *Neuron, 74,* 41–48.

Rubenstein, J. L. R., & Merzenich, M. M. (2003). Model of autism: Increased ratio of excitation/inhibition in key neural systems. *Genes, Brain and Behavior, 2,* 255–267.

Schizophrenia Working Group of the Psychiatric Genomics Consortium. (2014). Biological insights from 108 schizophrenia-associated genetic loci. *Nature, 511,* 421–427.

Sellers, K., Raval, P., & Srivastava, D. P. (2015). Molecular signature of rapid estrogen regulation of synaptic connectivity and cognition. *Frontiers of Neuroendocrinology, 36,* 72–89.

Seybold, B. A., Stanco, A., Cho, K., Potter, G., Kim, C., Sohal, V.,…Schreiner, C. E. (2012). Chronic reduction in inhibition reduces receptive field size in mouse auditory cortex. *Proceedings of the National Academy of Sciences, USA, 109,* 13829–13834.

Shimamura, K., & Rubenstein, J. L. (1997). Inductive interactions direct early regionalization of the mouse forebrain. *Development, 124,* 2709–2718.

Skucas, V. A., Duffy, A. M., Harte-Hargrove, L. C., Magagna-Poveda, A., Radman, T., Chakraborty, G.,…Scharfman, H. E. (2013). Testosterone depletion in adult male rats increases mossy fiber transmission, LTP, and sprouting in area CA3 of hippocampus. *Journal of Neuroscience, 33,* 2338–2355.

Storm, E. E., Garel, S., Borello, U., Hebert, J. M., Martinez, S., McConnell, S. K.,…Rubenstein, J. L. (2006). Dose-dependent functions of Fgf8 in regulating telencephalic patterning centers. *Development, 133,* 1831–1844.

Sun, T., & Walsh, C. A. (2006). Molecular approaches to brain asymmetry and handedness. *Nature Reviews Neuroscience, 7,* 655–662.

Upton, K. R., Gerhardt, D. J., Jesuadian, J. S., Richardson, S. R., Sanchez-Luque, F. J., Bodea, G. O.,…Faulkner, G. J. (2015). Ubiquitous L1 mosaicism in hippocampal neurons. *Cell, 161,* 228–239.

Visel, A., Taher, L., Girgis, H., May, D., Golonzhka, O., Hoch, R.,…Rubenstein, J. L. R. (2013). A high-resolution enhancer atlas of the developing telencephalon. *Cell, 152,* 895–908.

Yang, C. F., Chiang, M. C., Gray, D. C., Prabhakaran, M., Alvarado, M., Juntti, S. A.,…Shah, N. M. (2013). Sexually dimorphic neurons in the ventromedial hypothalamus govern mating in both sexes and aggression in males. *Cell, 153,* 896–909.

CHAPTER 4

Sex Differences in the Geschwind-Galaburda Hypothesis of Cerebral Lateralization

The Whole Body Perspective Revisited

Geert J. de Vries

SUMMARY

The role of sexual differentiation is a central theme of the Geschwind-Galaburda Hypothesis (GGH) of cerebral lateralization. The prediction that tissues involved in lateralized function and related disorders such as dyslexia are sexually dimorphic has been solidly confirmed with many examples of sex differences in gene expression, brain neurochemistry and morphology, and functional connectivity in lateralized structures. These differences are caused by sex differences in circulating hormones during development and in adulthood as well as direct actions of sex chromosomal genes. The occurrence and even the extent of some of these differences depend on the conditions under which individuals grow up. An important observation made by Geschwind and Galaburda was that sex differences in lateralization and the incidence of related disorders correlate with differences in peripheral systems such as the immune system. We now know that the latter differences may, in fact, contribute to neural sex difference. Therefore, it is essential to consider sex differences in the brain versus a whole body perspective. Doing so leads to the conclusion that sex differences in the brain do not necessarily cause differences in overt brain function and behavior, but may also do the exact opposite, promoting sex similarity by compensating for other differences, which would cause undesirable differences if left unchecked. This implies that certain cortically controlled functions and behaviors depend on sexually dimorphic neural circuitry, which suggests that the vulnerabilities for disorders in such function and behaviors are likely to differ as well.

INTRODUCTION

In the three landmark papers that together laid out the GGH regarding the lateralization of the brain and, by extension, the etiology of related learning disorders (LDs) such as dyslexia, Geschwind and Galaburda (1985a, b, c) argued that

> while genes contribute importantly, many influences that lie outside of the gene pool of the fetus can alter lateralization patterns. The most powerful factors are variations in the chemical environment in fetal life

> and to a lesser extent in infancy and early childhood. (Geschwind and Galaburda 1985a; p. 431)

Sexual differentiation and testosterone's role in this process were recurrent themes as they postulated, "on the basis of the higher frequency of sinistrality and of learning disorders in men, that this influence is related to male sex, e.g., testosterone or some factor related to it" (p. 431).

Rather than offering a simple hypothesis, the GGH papers charted a fertile area for further research on the origin of lateralization and related disorders. More recently, Galaburda offered a minimalist, and therefore more testable, version of the hypothesis: "Intrauterine testosterone causes variations in lateralizations (handedness and language), brain asymmetry, and contributes to learning disabilities and learning superiority, immunological and other disorders, and male–female differences" (personal communication, June 23, 2016). Although early claims of sex differences in the lateralization of the planum temporale (Wada, Clarke, & Hamm, 1975; Wada & Davis, 1977) could not be confirmed in later studies (Sommer, Aleman, Somers, Boks, & Kahn, 2008), the sexually dimorphic nature of the cortex has now been firmly established. For example, connectivity and functional MRI studies indicate that areas involved in language functions show anatomical as well as functional differences (see Chapters 7, 8, and 15). In addition, the molecular makeup of the brain, including human cortical cells, is dimorphic, with hundreds to thousands of genes showing sex differences in expression and epigenetic marks, some preceding gonadal differentiation (e.g., Dewing, Shi, Horvath, & Vilain, 2003; Ghahramani et al., 2014; Xu et al., 2014). So male and female brains clearly run by a different cellular and molecular script. This chapter discusses views on how such differences arise, the relation between such differences and function, and the implications of these sex differences for the vulnerability for behavior disorders and learning disabilities.

CAUSES OF SEX DIFFERENCES

The GGH argued that developmental differences in testosterone levels, and perhaps the presence or absence of the H-Y antigen, are the main factors driving sexual differentiation of the body, including the brain (Geschwind & Galaburda, 1985a). We now know that sex differences in organ systems throughout the body are not only based on hormone-dependent and independent actions of sex chromosomes but also on environmental influences. For example, numerous studies confirmed that early differences in gonadal hormones cause anatomical, biochemical, and functional differences by directing the course of development (programming or organizational effects); in addition, later differences in circulating gonadal hormones cause additional sex differences (acute or activational effects) (Forger, De Vries, & Breedlove, 2015). Although testosterone still stands as the most important programming factor, nonsteroidal factors secreted by the developing testis (e.g., Mullerian inhibiting hormone [MIH]) may play a role as well. MIH not only has a differentiating effect on telencephalic and cerebellar tissue (Wittmann & McLennan, 2011, 2013a, b), but it may also contribute to the male bias in autism (Pankhurst & McLennan, 2012). Although nonsteroidal testicular secretions have not been linked with LDs such as dyslexia, it would be interesting to explore this link.

Testosterone influences sexual differentiation in humans mainly by signaling via the androgen receptor (Forger et al., 2015; Hines, Ahmed, & Hughes, 2003). Measuring the ratio of the length of the second and fourth digits (2D:4D), which is larger in females than in males from fetal life onward (Breedlove, 2010), is one way to indirectly test whether a sexually dimorphic human trait is influenced by prenatal levels of testosterone. At least two studies reported that 2D:4D is lower in left- than in right-handed males (Manning & Peters, 2009; Stoyanov, Marinov, & Pashalieva, 2009), suggesting that sex differences in handedness depend at least in part on programming effects of testosterone. Similar studies did not find a relationship between 2D:4D and dyslexia (Boets, De Smedt, Wouters, Lemay, & Ghesquiere, 2007; van Gelder, Tijms, & Hoeks, 2005). This discrepancy between left-handedness and dyslexia suggests that factors other than testosterone play a role in sex differences in dyslexia. Boets and colleagues, however, did not find a relationship between 2D:4D and handedness. A more direct test to assess the role of androgen signaling might be to study dyslexia in XY females with androgen insensitivity syndrome (cf. Hines et al., 2003). These XY females incidentally have a female 2D:4D (Berenbaum, Bryk, Nowak, Quigley, & Moffat, 2009). If testosterone plays a role in dyslexia, then one would predict the disorder to be much less common in these XY females than in XY males.

When the GGH was originally proposed, the field of sexual differentiation was still heavily dominated by Alfred Jost's proposal in the 1940s that the testes are crucial for the development of the male phenotype, and the body develops in a female direction without them (Jost, 1970). Additional findings have put the universality of this doctrine into question and have suggested direct sexually dimorphic effects of sex chromosomal genes on nongonadal tissue, including the brain (Arnold, 1996). Such effects can be caused by Y-specific genes (e.g., those coding for the H-Y antigen) and X chromosomal dosage effects (Arnold, 2009). The GGH appears to have foreseen this as it stated

> One should not assume that H-Y antigen exerts all its effects only indirectly through the formation of testes and the later production of T: this antigen is expressed on other organs, e.g., the thymus gland, an essential component of the developing immune system. At conception the female zygote carries twice as many X-linked genes as the male. The X chromosome plays an important role in effects of T since it contains the Tfm locus, one of the genes that controls sensitivity to this hormone. (Geschwind & Galaburda, 1985b, p. 524)

We now know that the Sry gene, which is present on the Y chromosome, is necessary for testis development instead of the H-Y antigen (Koopman, Gubbay, Vivian, Goodfellow, & Lovellbadge, 1991). Nevertheless, we can test the role of this antigen in the differentiation of the brain more directly now that the genes that code for the H-Y antigen have been identified (Scott et al., 1995; Wang et al., 1995).

To differentiate between direct effects of the sex chromosomal complement (XX *vs.* XY) and gonadal hormones, our research team and others introduced the four core genotype mice model, which consists of XX and XY mice of either sex that are generated by crossing XX females with XY Sry males (Arnold, 2004; De Vries et al., 2002). The latter lack the Sry gene on the Y chromosome but have a Sry transgene inserted on an autosomal chromosome, which thereby takes over the sex-determining role of the Y chromosome. Comparing XX and XY mice within sex (defined by having testes or ovaries) has so far revealed numerous differences in neural as well as nonneural tissue that seem to be caused directly by differences in sex chromosomal complement (Arnold, Chen, Link, Itoh, & Reue, 2013). Use of these mice and related mouse models that can decipher even more precisely which aspects of sex chromosomes direct sex bias (e.g., dosage of X chromosomal genes) (Arnold, 2014) will be helpful in studying direct effects of sex chromosomal genes on the development of neural circuits involved in cognitive functions. For example, that such effects will be found follows from cortical Sry expression, which is lacking in XX females (Dewing et al., 2006). Imaging research in humans with sex chromosomal anomalies also suggests direct effects of sex chromosomal genes. For example, women with Turner syndrome, who have only one X chromosome and no Y chromosome, or males with Klinefelter syndrome, who have two X chromosomes plus a Y chromosome, differ from sex-matched controls (XX women and XY men) in gray matter volume of specific cortical areas corresponding to the number of X chromosomes (Hong et al., 2014). Although the functions of these sex differences are unknown, they indicate differences in physiology.

FUNCTION OF SEX DIFFERENCES

In general, identifying the functional significance of neural sex differences is surprisingly difficult. Initial research aimed at linking differences in reproductive physiology and behavior to differences in brain structure (De Vries, 2004). The finding of numerous sex differences in areas implicated in sexually dimorphic functions has reinforced the idea that sex differences in the brain beget sex differences in behavior. Our research team realized that the opposite is probably true as well while studying the vasopressin innervation of the prairie vole brain, a species that shows remarkably little sex differences in behavior but has one of the biggest differences in vasopressin innervation among vertebrates, with males having much denser innervation than females. Our results suggested that this particular difference may promote similarities in behavior. We therefore proposed that, in addition to causing functional and behavioral differences, sex differences in the brain may in some cases compensate for other differences, which if left unchecked may cause undesirable sex differences (De Vries & Boyle, 1998).

This dual function idea appears valid for every level of organization and, in fact, may be even easier to grasp at the molecular level. For example, one of the most pervasive sex differences is the inactivation of one of the X chromosomes in every cell in the female body, which is completely absent in male cells (Lyon, 1961). An obvious reason for this is that some genes on the X chromosome control regular, not necessarily sexually dimorphic, household functions and, thus, may cause undesirable effects if expressed in double dosage. Therefore, the most likely explanation is that this difference ensures similarity in function. Comparable types of compensation are probably necessary at higher levels of organization as well. For example, it is unlikely that there are brain circuits dedicated only to controlling functions such as sexual behavior. More likely, such functions are generated by distributed networks that share nodes with circuits controlling nondimorphic behaviors. Sex differences in shared nodes will have consequences for both circuits, and adaptations may be needed to have all circuits function optimally, whether they generate dimorphic or nondimorphic functions (De Vries, 2004).

WHOLE BODY PERSPECTIVE OF SEX DIFFERENCES

Of course, sexually dimorphic circuits do not operate in a vacuum. They communicate with and within bodies that themselves show differences that may require adaptations for optimal function. Geschwind and Galaburda (1985b) were aware of such differences and pointed out that "males and females differ in almost every aspect of visible body structure, but there are differences that transcend obvious structural features, e.g., gender-related variations in the immune system in responsiveness to drugs and toxins and in metabolism" (p. 524).

They also mentioned that the same conditions that lead to sex differences in lateralization and incidence of dyslexia may lead to differences in immune function: "The factors that modify cerebral dominance also influence the development of many other systems, e.g., the organs involved in immune response" (Geschwind & Galaburda, 1985a, p. 431). Now we know that in many cases such differences may be more than just correlational. In a recent review, we identified multiple sex differences in organ systems that trigger differences in neural structure, function, and dysfunction (De Vries & Forger, 2015) (see Figure 4.1).

For example, the same immune stimuli cause different immune responses in males and females (Mirandola et al., 2015), which have been related to the higher incidence of autoimmune diseases and increased

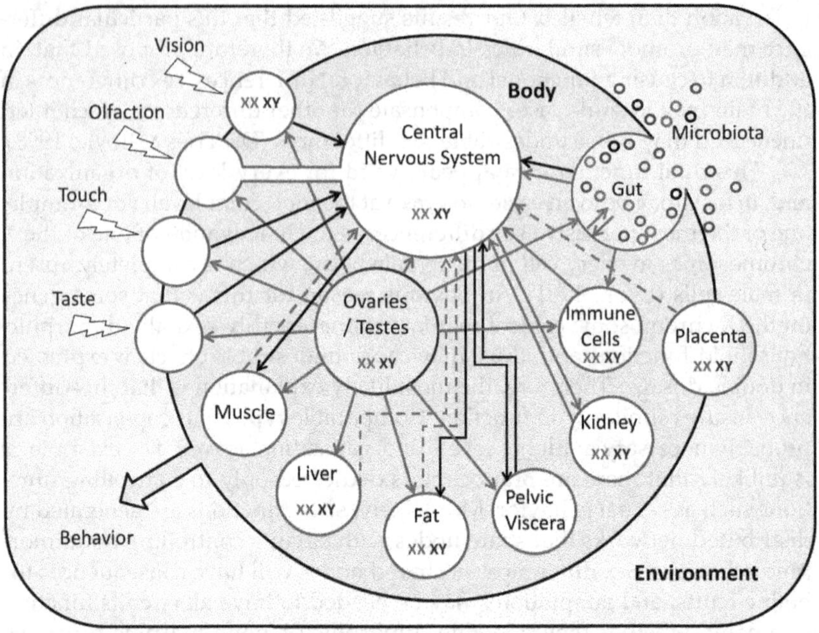

Figure 4.1. Sex differences in peripheral influences on the central nervous system. The CNS is embedded in a sexually differentiated body that is embedded in an environment, which may interact with the body in a manner that varies by sex. This diagram represents interactions between the central nervous system and sex differences elsewhere in the body that are discussed in this paper; other interactions undoubtedly occur. *Solid arrows* indicate a sex influence from one organ on another. *Dashed arrows* indicate an influence inferred from circumstantial evidence, but not yet demonstrated. *Black arrows* indicate neural communication; *red arrows* indicate humoral communication. *"XX XY"* indicates organs in which sex chromosome complement has a demonstrated effect; in most cases, it is not known whether the effect is mediated within that organ or indirectly via effects on other organs. The *small colored circles* in the *upper right* are the many species of microorganisms living commensally in our gut or on our skin. (Copyright © De Vries and Forger. 2015; published originally by BioMed Central.) A full-color version of this figure is available online at http://www.brookespublishing.com/galaburda/materials.

responses to infections in women (Beeson, 1994; Klein, 2012; Klein, Marriott, & Fish, 2015). Differences in immune activation and level of inflammation during critical periods of development have, in turn, been linked to sex-specific effects on brain functions such as learning and memory, locomotion, emotional regulation, and social behavior, thereby often blunting or increasing differences in these functions (Bale, 2009; Taylor et al., 2012).

A related and emerging focus are the physiological and immune responses to the microbiota, the collection of microorganisms that live symbiotically on and within our bodies. Differences in species composition of the microbiota (e.g., caused by illness, diet, or immune disorders) can influence the absorption of nutrients, the secretion of lymphokines and cytokines by intestinal tissue, and even the absorption of chemicals secreted by bacteria, all of which can influence the physiology of different tissues, including the brain (Cryan & Dinan, 2012). The gut microbiota in mice becomes sexually dimorphic after puberty, reflecting sex differences in adult hormone levels. In turn, the microbiota influence circulating testosterone levels in males, and transfer of gut microbiota from adult males to immature females alters the hormone levels as well as autoimmune responses of the recipients (Markle et al., 2013). Although consequences of these differences for sexual differentiation of the brain have not been studied, work suggests that the microbiota has sexually dimorphic programming as well as acute effects on social behavior: germ-free Swiss Webster mice are less sociable and do not show the typical preference for investigating novel mice (Desbonnet, Clarke, Shanahan, Dinan, & Cryan, 2014). Colonizing germ-free mice with microbial species at day 21 (prepuberty) only partially normalized social responses in adulthood (Desbonnet et al., 2014), suggesting that the microbiota must be present even earlier for normal social behavior. Interestingly, these effects are stronger in males than in females, suggesting that the microbiota can influence sexual differentiation.

One explanation for sex-specific effects of immune challenges as well as microbiota composition is that the brain itself uses elements of immune or inflammatory signaling in the process of sexual differentiation of brain and behavior (e.g., Amateau & McCarthy, 2004; Bale, 2009; Petersen, Krishnan, Aggison, Intlekofer, & Moura, 2012). Because this is also true for cortical circuits (e.g., Adelson et al., 2016; Macht, 2016), early immune challenges may also influence sex differences in lateralization of cortical function.

CONSEQUENCES OF SEX DIFFERENCES FOR DISEASE

The dual function hypothesis for sex differences in the brain previously mentioned is eminently testable. If a sexual dimorphism in the brain causes a difference in physiology or behavior, then one would predict that reducing the impact of that dimorphism (e.g., by administering antagonists to block a sexually dimorphic neurotransmitter system) will reduce functional differences; if a dimorphism compensates for other differences, then reducing its impact would create a difference in physiology and behavior where there was not one before. Such experiments have been done for the previously mentioned sex difference in vasopressin innervation (De Vries & Södersten, 2009). For example, vasopressin receptor knockout mice have a behavioral phenotype in males (lower anxiety in knockouts), but there is no effect in females (Bielsky, Hu, Ren, Terwilliger, & Young, 2005; Bielsky, Hu, & Young, 2005), which is what one would predict for a system that is more important for certain functions in one sex than in the other.

Interestingly, nature may have done similar experiments. For example, a specific polymorphism in the promoter region of the V1a gene correlates with differences in pairbonding in men but not in women; men with that polymorphism are less likely to be married and, if they are married, then they are more likely to have had serious marital discord (Walum et al.,

2008). Because the same polymorphism has also been linked to disorders such as autism (Kim et al., 2002; Wassink et al., 2004; Yirmiya et al., 2006), it could potentially contribute to sex differences in these disorders. Without question, the human cortex differs between males and females in gene expression as well as in anatomy and functional connectivity (see Chapters 7, 8, and 15). In principle, these sex differences may cause, as well as prevent, sex differences in function. This implies that certain cortically controlled behaviors and functions, whether sexually differentiated, depend on sexually dimorphic neural circuitry. If so, the vulnerabilities for disorders in those behaviors and functions likely differ as well.

REFERENCES

Adelson, J. D., Sapp, R. W., Brott, B. K., Lee, H., Miyamichi, K., Luo, L. Q.,...Shatz, C. J. (2016). Developmental sculpting of intracortical circuits by mhc class i h2-db and h2-kb. *Cerebral Cortex, 26*(4), 1453–1463.

Amateau, S. K., & McCarthy, M. M. (2004). Induction of pge(2) by estradiol mediates developmental masculinization of sex behavior. *Nature Neuroscience, 7*(6), 643–650.

Arnold, A. P. (1996). Genetically triggered sexual differentiation of brain and behavior. *Hormones and Behavior, 30*(4), 495–505.

Arnold, A. P. (2004). Sex chromosomes and brain gender. *Nature Reviews Neuroscience, 5*(9), 701–708.

Arnold, A. P. (2009). The organizational-activational hypothesis as the foundation for a unified theory of sexual differentiation of all mammalian tissues. *Hormones and Behavior, 55*(5), 570–578.

Arnold, A. P. (2014). Conceptual frameworks and mouse models for studying sex differences in physiology and disease: Why compensation changes the game. *Experimental Neurology, 259*, 2–9.

Arnold, A. P., Chen, X. Q., Link, J. C., Itoh, Y., & Reue, K. (2013). Cell-autonomous sex determination outside of the gonad. *Developmental Dynamics, 242*(4), 371–379.

Bale, T. L. (2009). Neuroendocrine and immune influences on the cns: It's a matter of sex. *Neuron, 64*(1), 13–16.

Beeson, P. B. (1994). Age and sex associations of 40 autoimmune-diseases. *American Journal of Medicine, 96*(5), 457–462.

Berenbaum, S. A., Bryk, K. K., Nowak, N., Quigley, C. A., & Moffat, S. (2009). Fingers as a marker of prenatal androgen exposure. *Endocrinology, 150*(11), 5119–5124.

Bielsky, I. F., Hu, S. B., Ren, X., Terwilliger, E. F., & Young, L. J. (2005). The v1a vasopressin receptor is necessary and sufficient for normal social recognition: A gene replacement study. *Neuron, 47*(4), 503–513.

Bielsky, I. F., Hu, S. B., & Young, L. J. (2005). Sexual dimorphism in the vasopressin system: Lack of an altered behavioral phenotype in female v1a receptor knockout mice. *Behavioural Brain Research, 164*(1), 132–136.

Boets, B., De Smedt, B., Wouters, J., Lemay, K., & Ghesquiere, P. (2007). No relation between 2d:4d fetal testosterone marker and dyslexia. *Neuroreport, 18*(14), 1487–1491.

Breedlove, S. M. (2010). Minireview: Organizational hypothesis: Instances of the fingerpost. *Endocrinology, 151*(9), 4116–4122.

Cryan, J. F., & Dinan, T. G. (2012). Mind-altering microorganisms: The impact of the gut microbiota on brain and behaviour. *Nature Reviews Neuroscience, 13*(10), 701–712.

De Vries, G. J. (2004). Minireview: Sex differences in adult and developing brains: Compensation, compensation, compensation. *Endocrinology, 145*(3), 1063–1068.

De Vries, G. J., & Boyle, P. A. (1998). Double duty for sex differences in the brain. *Behavioral Brain Research, 92*(2), 205–213.

De Vries, G. J., & Forger, N. G. (2015). Sex differences in the brain: A whole body perspective. *Biology of Sex Differences, 6*, 15.

De Vries, G. J., Rissman, E. F., Simerly, R. B., Yang, L. Y., Scordalakes, E. M., Auger, C. J.,… Arnold, A. P. (2002). A model system for study of sex chromosome effects on sexually dimorphic neural and behavioral traits. *Journal of Neuroscience, 22*(20), 9005–9014.
De Vries, G. J., & Södersten, P. (2009). Sex differences in the brain: The relation between structure and function. *Hormones and Behavior, 55*(5), 589–596.
Desbonnet, L., Clarke, G., Shanahan, F., Dinan, T. G., & Cryan, J. F. (2014). Microbiota is essential for social development in the mouse. *Molecular Psychiatry, 19*(2), 146–148.
Dewing, P., Chiang, C. W. K., Sinchak, K., Sim, H., Fernagut, P. O., Kelly, S.,…Vilain, E. (2006). Direct regulation of adult brain function by the male-specific factor sry. *Current Biology, 16*(4), 415–420.
Dewing, P., Shi, T., Horvath, S., & Vilain, E. (2003). Sexually dimorphic gene expression in mouse brain precedes gonadal differentiation. *Molecular Brain Research, 118*(1-2), 82–90.
Forger, N. G., De Vries, G. J., & Breedlove, S. M. (2015). Sexual differentiation of brain and behavior. In T.M. Plant & A. J. Zeleznik (Eds.), *Knobil and Neill's physiology of reproduction* (4th ed., Vol. 2, pp. 2109–2155). New York, NY: Academic Press.
Geschwind, N., & Galaburda, A. M. (1985a). Cerebral lateralization. Biological mechanisms, associations, and pathology: I. A hypothesis and a program for research. *Archives of Neurology, 42*(5), 428–459.
Geschwind, N., & Galaburda, A. M. (1985b). Cerebral lateralization. Biological mechanisms, associations, and pathology: II. A hypothesis and a program for research. *Archives of Neurology, 42*(6), 521–552.
Geschwind, N., & Galaburda, A. M. (1985c). Cerebral lateralization. Biological mechanisms, associations, and pathology: III. A hypothesis and a program for research. *Archives of Neurology, 42*(7), 634–654.
Ghahramani, N. M., Ngun, T. C., Chen, P. Y., Tian, Y., Krishnan, S., Muir, S.,…Vilain, E. (2014). The effects of perinatal testosterone exposure on the DNA methylome of the mouse brain are late-emerging. *Biology of Sex Differences, 5*, 8.
Hines, M., Ahmed, S. F., & Hughes, I. A. (2003). Psychological outcomes and gender-related development in complete androgen insensitivity syndrome. *Archives of Sexual Behavior, 32*(2), 93–101.
Hong, D. S., Hoeft, F., Marzelli, M. J., Lepage, J. F., Roeltgen, D., Ross, J., & Reiss, A. L. (2014). Influence of the x-chromosome on neuroanatomy: Evidence from Turner and Klinefelter syndromes. *Journal of Neuroscience, 34*(10), 3509–3516.
Jost, A. (1970). Hormonal factors in the sex differentiation of the mammalian foetus. *Philosophical Transactions of the Royal Society of London Series B-Biological Sciences, 259*(828), 119–130.
Kim, S. J., Young, L. J., Gonen, D., Veenstra-VanderWeele, J., Courchesne, R., Courchesne, E.,…Insel, T. R. (2002). Transmission disequilibrium testing of arginine vasopressin receptor 1a (avpr1a) polymorphisms in autism. *Molecular Psychiatry, 7*(5), 503–507.
Klein, S. L. (2012). Sex influences immune responses to viruses, and efficacy of prophylaxis and treatments for viral diseases. *Bioessays, 34*(12), 1050–1059.
Klein, S. L., Marriott, I., & Fish, E. N. (2015). Sex-based differences in immune function and responses to vaccination. *Transactions of the Royal Society of Tropical Medicine and Hygiene, 109*(1), 9–15.
Koopman, P., Gubbay, J., Vivian, N., Goodfellow, P., & Lovellbadge, R. (1991). Male development of chromosomally female mice transgenic for sry. *Nature, 351*(6322), 117–121.
Lyon, M. F. (1961). Gene action in the x-chromosome of the mouse (mus musculus l.). *Nature, 190*(477), 372–373.
Macht, V. A. (2016). Neuro-immune interactions across development: A look at glutamate in the prefrontal cortex. *Neuroscience and Biobehavioral Reviews* 71, 267–280.
Manning, J. T., & Peters, M. (2009). Digit ratio (2d:4d) and hand preference for writing in the BBC internet study. *Laterality, 14*(5), 528–540.
Markle, J. G. M., Frank, D. N., Mortin-Toth, S., Robertson, C. E., Feazel, L. M., Rolle-Kampczyk, U.,…Danska, J. S. (2013). Sex differences in the gut microbiome drive hormone-dependent regulation of autoimmunity. *Science, 339*(6123), 1084–1088.
Mirandola, L., Wade, R., Verma, R., Pena, C., Hosiriluck, N., Figueroa, J. A.,…Chiriva-Internati, M. (2015). Sex-driven differences in immunological responses: Challenges and

opportunities for the immunotherapies of the third millennium. *International Reviews of Immunology, 34*(2), 134–142.

Pankhurst, M. W., & McLennan, I. S. (2012). Inhibin b and anti-mullerian hormone/mullerian-inhibiting substance may contribute to the male bias in autism. *Translational Psychiatry, 2*, e148.

Petersen, S. L., Krishnan, S., Aggison, L. K., Intlekofer, K. A., & Moura, P. J. (2012). Sexual differentiation of the gonadotropin surge release mechanism: A new role for the canonical nf kappa b signaling pathway. *Frontiers of Neuroendocrinology, 33*(1), 36–44.

Scott, D. M., Ehrmann, I. E., Ellis, P. S., Bishop, C. E., Agulnik, A. I., Simpson, E., & Mitchell, M. J. (1995). Identification of a mouse male-specific transplantation antigen, H-Y. *Nature, 376*(6542), 695–698.

Sommer, I. E., Aleman, A., Somers, M., Boks, M. P., & Kahn, R. S. (2008). Sex differences in handedness, asymmetry of the planum temporale and functional language lateralization. *Brain Research, 1206*, 76–88.

Stoyanov, Z., Marinov, M., & Pashalieva, I. (2009). Finger length ratio (2d:4d) in left- and right-handed males. *International Journal of Neuroscience, 119*(7), 1006–1013.

Taylor, P. V., Veenema, A. H., Paul, M. J., Bredewold, R., Isaacs, S., & De Vries, G. J. (2012). Sexually dimorphic effects of a prenatal immune challenge on social play and vasopressin expression in juvenile rats. *Biology of Sex Differences, 3*, doi:10.1186/2042-6410-3-15

van Gelder, M., Tijms, J., & Hoeks, J. (2005). Second to fourth digit ratio and dyslexia: No evidence for an association between reading disabilities and the 2d:4d ratio. *Developmental Medicine and Child Neurology, 47*(10), 718–719. doi:10.1017/s0012162205211477

Wada, J. A., Clarke, R., & Hamm, A. (1975). Cerebral hemispheric asymmetry in humans. Cortical speech zones in 100 adults and 100 infant brains. *Archives of Neurology, 32*(4), 239–246.

Wada, J. A., & Davis, A. E. (1977). Fundamental nature of human infant's brain asymmetry. *Journal of Neurological Sciences. Le Journal Canadien des Sciences Neurologiques, 4*(3), 203–207.

Walum, H., Westberg, L., Henningsson, S., Neiderhiser, J. M., Reiss, D., Igl, W.,…Lichtenstein, P. (2008). Genetic variation in the vasopressin receptor 1a gene (avpr1a) associates with pair-bonding behavior in humans. *Proceedings of the National Academy of Sciences USA, 105*(37), 14153–14156. doi:10.1073/pnas.0803081105

Wang, W., Meadows, L. R., Denhaan, J. M. M., Sherman, N. E., Chen, Y., Blokland, E.,…Engelhard, V. H. (1995). Human h-y: A male-specific histocompatibility antigen derived from the smcy protein. *Science, 269*(5230), 1588–1590.

Wassink, T. H., Piven, J., Vieland, V. J., Pietila, J., Goedken, R. J., Folstein, S. E., & Sheffield, V. C. (2004). Examination of AVPR1a as an autism susceptibility gene. *Molecular Psychiatry, 9*(10), 968–972.

Wittmann, W., & McLennan, I. S. (2011). The male bias in the number of purkinje cells and the size of the murine cerebellum may require mullerian inhibiting substance/anti-mullerian hormone. *Journal of Neuroendocrinology, 23*(9), 831–838.

Wittmann, W., & McLennan, I. S. (2013a). Anti-mullerian hormone may regulate the number of calbindin-positive neurons in the sexually dimorphic nucleus of the preoptic area of male mice. *Biology of Sex Differences, 4*(1), 18.

Wittmann, W., & McLennan, I. S. (2013b). The bed nucleus of the stria terminalis has developmental and adult forms in mice, with the male bias in the developmental form being dependent on testicular amh. *Hormones and Behavior, 64*(4), 605–610.

Xu, H. Q., Wang, F., Liu, Y. W., Yu, Y. Q., Gelernter, J., & Zhang, H. P. (2014). Sex-biased methylome and transcriptome in human prefrontal cortex. *Human Molecular Genetics, 23*(5), 1260–1270.

Yirmiya, N., Rosenberg, C., Levi, S., Salomon, S., Shulman, C., Nemanov, L.,…Ebstein, R. P. (2006). Association between the arginine vasopressin 1a receptor (avpr1a) gene and autism in a family-based study: Mediation by socialization skills. *Molecular Psychiatry, 11*(5), 488–494.

CHAPTER 5

Can Prenatal Maternal Distress Predict Child Cerebral Laterality?

Recent Findings and Implications for the Geschwind-Galaburda Hypothesis

Thomas G. O'Connor, Emily S. Barrett, and Ana Vallejo Sefair

SUMMARY

Research findings collected over the past two decades, and particularly in recent years, indicate that prenatal maternal distress (PMD) is reliably associated with a range of neurodevelopmental outcomes in the child. This chapter examines this evidence in the context of the Geschwind-Galaburda Hypothesis (GGH; Geschwind & Galaburda, 1985a, b, c) on cerebral laterality. We consider the compatibilities and disjunctions between the research on PMD and the GGH in terms of behavioral phenotypes in the child, mediating mechanisms, and possible clinical application. We conclude that although most of the research on PMD did not specifically target the GGH, there are, nonetheless, striking areas of confluence and a considerable basis for further specific areas of investigation.

INTRODUCTION

Prenatal chemical exposures that may partly account for the cerebral lateralization patterns and associated conditions in the child is one principal component of the GGH (Geschwind & Galaburda, 1985a). "Factors that modify cerebral dominancy also influence the development of many other systems, e.g., the organs involved in the immune response" (Geschwind & Galaburda, 1985a, p. 431) is another component of the GGH. This chapter aims to examine how research on prenatal maternal psychological distress parallels research on the GGH and may be one factor accounting for lateralization that is central to the GGH.

A PRIMER OF THE PRENATAL MATERNAL DISTRESS PARADIGM

Research on PMD now constitutes a substantial research program that has captivated attention from diverse scientific disciplines in many countries. The notion that maternal mood in pregnancy may cause or foretell something about the child has a long history in folk psychology and mythology, but the scientific tackling of this question for understanding human health

Support for research reported in this paper was provided by National Institutes of Health grants MH073019, MH073842, and HD083369 and the March of Dimes Foundation.

derives from decades of experimental animal work (e.g., Ader & Plaut, 1968). Experimental studies have extended that early animal work considerably by identifying the biological mechanisms that could account for the growing and remarkably diverse outcomes in the offspring associated with prenatal stress (Coe, Lubach, & Karaszewski, 1999; Darnaudery & Maccari, 2008; Henry, Kabbaj, Simon, Le Moal, & Maccari, 1994; van Batenburg-Eddes et al., 2009).

As will be noted in a later section, many different constructs indicating maternal psychological distress have been used in the research literature, including anxiety, depression, and objective and subjective measures of stress. The term *distress* is used throughout the chapter as a broad term to encompass the broad literature; more specific terms are used when referring to individual studies or particular findings.

Human research on PMD takes its lead from fetal programming (Barker, 1993; Hanson & Gluckman, 2011) and predictive adaptive response (Gluckman & Hanson, 2005) models. In brief, these models hypothesize that in utero exposures instigate an adaptive response in the fetus/child and that these programmed responses are carried forward in development with persisting (and putatively adaptive) effects on behavior and biology. The exposure variable—prenatal maternal psychological distress—is presumed to be a proxy for multiple biological changes in the maternal endocrine, hormone, and immune systems that are responsive to stress and may be directly or indirectly communicated to the fetus. How maternal stress-linked biology may gain access to the fetus and under what conditions and through what mechanisms is now a focus of concerted work ranging from intensive behavioral research (Entringer, Buss, Andersen, Chicz-Demet, & Wadhwa, 2011) to analyses of placental gene expression (e.g., O'Donnell et al., 2012).

Empirical evidence implying a robust association between PMD and child outcomes is substantial; numerous reviews are available (Entringer, Buss, & Wadhwa, 2010; Glover, O'Connor, & O'Donnell, 2010; Talge, Neal, & Glover, 2007). Next, we discuss findings with special relevance to the GGH. We offer two caveats at the outset. One is that PMD indexes multiple overlapping and confounded mechanisms, ranging from glucocorticoid exposure to behavioral health decisions; the implication is that the effects of PMD will not be traced to a singular cause. The second concerns the risk phenotype. We use the term *maternal distress* to cover an array of constructs and measures that appear in research studies, such as acute stress or major life events, continuous measures of psychological symptoms, pregnancy-related anxiety, psychiatric diagnoses, trauma symptoms, or traumatic exposures.

PRENATAL MATERNAL DISTRESS, LATERALITY, AND RELATED OUTCOMES

As with the GGH on cerebral laterality, the PMD hypotheses make the explicit prediction that prenatal exposures may have a particular influence on child development (e.g., distinct from postnatal exposures and genetic confounds). We examine several kinds of evidence to consider whether PMD may represent one possible mechanism underlying cerebral laterality.

The most direct evidence that PMD may be associated with laterality comes from the few studies that assess handedness, perhaps the only laterality measure to attract attention in several studies. Analyses of more than 7,000 parents and children from the Avon Longitudinal Study of Parents and Children (ALSPAC) cohort indicated that prenatal maternal anxiety at 18 weeks gestation was associated with mixed (not right and not left) handedness in 42-month-old children (odds ratio, OR = 1.23) (Glover, O'Connor, Heron, & Golding, 2004). The observation is similar to that reported in a Danish cohort of more than 800 3-year-old children (Obel, Hedegaard, Henriksen, Secher, & Olsen, 2003). A third study of approximately 100 6-year-olds found more mixed evidence (Gutteling, de Weerth, & Buitelaar, 2007). It is noteworthy that, to date, there are no robust sex differences in this prediction, despite some evidence that PMD may alter laterality.

A second tier of evidence includes studies linking PMD to child neurodevelopmental outcomes associated with atypical laterality, such as attention-deficit/hyperactivity disorder (Hale et al., 2009) and autism (Lindell & Hudry, 2013); of course, many of these neurodevelopmental outcomes also show sizable sex differences, which further underscores their (indirect) relevance for research on laterality, given the prominent sex differences in laterality. Research findings indicate robust associations between PMD and attention problems in the child, with evidence persisting into adolescence (Li, Olsen, Vestergaard, & Obel, 2010; O'Donnell, Glover, Barker, & O'Connor, 2014; van den Bergh et al., 2006). Effect sizes, which are imperative for understanding clinical significance, are not routinely formally calculated (although they can usually be calculated from reported information), but some reports suggest as much as a doubling of the risk for significant attention problems from PMD. The broader neurodevelopmental phenotype that includes executive function, language ability, and general cognitive ability has also been extensively studied, with considerable evidence showing a link with PMD (e.g., Buss, Davis, Hobel, & Sandman, 2011; Laplante et al., 2004).

More severe clinical conditions that are linked with laterality and show sex differences have also been analyzed in research on PMD, but the evidence is more limited and not yet robust. One such example is autism, for which the available evidence of an association with PMD is mixed, with both supportive (Kinney, Munir, Crowley, & Miller, 2008) and nonsupportive (Li et al., 2009) findings reported. There are intriguing findings concerning PMD and schizophrenia (Khashan et al., 2008) and schizophrenia-associated phenotypes (King et al., 2009), although that literature is also limited. Other developmental disabilities and outcomes of particular relevance to cerebral laterality, such as dyslexia, have received too little research attention in this area to review.

Sex differences are a major consideration in research on cerebral laterality, but the findings on sex differences in response to prenatal distress in humans is far from clear. To be sure, some have been reported in human studies, and sex differences are omnipresent in animal studies; however, the counterweight is from large-scale human studies that found no evidence of sex differences, even for those neurodevelopmental outcomes linked with laterality that show sex differences in prevalence, such as attention problems (O'Donnell et al., 2014).

A further kind of evidence of obvious significance for understanding PMD on brain development, including cerebral laterality, derives from the small but rapidly growing literature assessing brain images in the child. Essentially no studies that specifically targeted PMD as the key exposure variable set out to test the GGH, but several findings are worth noting. Buss and colleagues (Buss et al., 2011) studied 30 boys and 35 girls approximately 7 years of age and reported that higher maternal cortisol measured at 15 weeks gestation was associated with greater right amygdala volume in girls only; they did not find volumetric differences in the hippocampus. Sarkar and colleagues (Sarkar et al., 2014) conducted a diffusion tensor imaging (DTI) study of 22 children who were an average of 7 years old and reported that prenatal stressful life events were associated with white matter microstructural organization of the right uncinate fasciculus; a similar trend was found for the left uncinate fasciculus (too few participants were studied to examine child sex as a moderator). Interestingly, postnatal life stress did not have comparable prediction. Another DTI study of 54 neonates found that prenatal maternal anxiety was associated with several measures of brain microstructure, including lower fractional anisotropy in right insular cortex, right uncinate fasciculus, and right dorsolateral prefrontal cortex; bilateral effects were also observed (Rifkin-Graboi et al., 2015). These studies point to the growing interest in incorporating brain imaging in studies of prenatal maternal stress and suggest that the next several years will yield a great deal of data that will require interpretation and orientation to the cerebral laterality hypotheses. It will be challenging to integrate imaging findings across different techniques and resolve the likely inevitable empirical inconsistencies.

An additional tier of evidence has at least marginal relevance to PMD and laterality. Jones, Osmond, Godfrey, and Phillips (2011) studied cerebral laterality based on contrasting tympanic membrane temperatures (TMT) in 8- to 9-year-old children and found that low birth weight, as an index of an adverse prenatal environment, was associated with TMT laterality. Specifically, fetal growth was positively associated with right TMT in response to a mild stressor, indicating reduced right hemisphere activity. These results may be indirectly relevant to PMD because of the link between prenatal stress and obstetric outcomes; the potential of the TMT as a measure of cerebral laterality is worth highlighting.

PRENATAL MATERNAL STRESS AND PRENATAL HORMONE EXPOSURE

Prenatal testosterone exposure is central in GGH research because it may account, in part, for the observed cerebral laterality and sex differences found in neurodevelopment. Empirical evidence comes from a variety of research designs. One type of study indicating the involvement of prenatal testosterone in cerebral lateralization contrasts outcomes in girls from opposite-sex and same-sex twin pairs. A study of 53 girls from same-sex dizygotic pairs, 67 girls from opposite-sex pairs, and 67 boys from opposite-sex twin pairs (Cohen-Bendahan, Buitelaar, van Goozen, & Cohen-Kettenis, 2004) found a more masculine pattern of right ear dominance and cerebral lateralization in a dichotic listening task (a leading measure of functional cerebral lateralization) in girls from opposite-sex pairs than same-sex twin pairs. Furthermore, testosterone assessed at the time of the dichotic listening task, at approximately 11 years of age, was not associated with performance. Also, girls in same-sex twin pairs had the highest mean level of free testosterone. The implication is that increased prenatal exposure to testosterone in the opposite-sex pairs caused the greater masculinized pattern. Prenatal testosterone may also be a causal factor in neurodevelopmental outcomes associated with cerebral laterality and sex differences. Autism has received a good deal of attention in this area (Auyeung et al., 2012), although an analysis indicated that autism-related measures were not associated with amniotic fluid testosterone and were not greater in girls according to whether the mother had congenital adrenal hyperplasia (Kung et al., 2016). This issue clearly requires further exploration.

The hypothalamic pituitary adrenal (HPA) axis, and glucocorticoids in particular, is one dominant focus in research on how PMD may shape offspring development in animal literature, in which the evidence base is substantial (Darnaudery & Maccari, 2008; Matthews, 2000; Owen, Andrews, & Matthews, 2005), and in human research, in which the evidence base is modest but growing (Bergman, Sarkar, Glover, & O'Connor, 2010; O'Donnell et al., 2012; Seckl & Meaney, 2004; Wadhwa, Dunkel-Schetter, Chicz-DeMet, Porto, & Sandman, 1996). More important, however, the animal and human literatures also indicate that there may be an overfocus on the HPA axis and a significant underappreciation for the role of sex steroids in the causal model (O'Connor & Barrett, 2014). For example, the animal literature indicated that restoring testosterone levels (decreased in response to prenatal stress) eliminated behavioral differences associated with prenatal stress in male offspring (Kapoor & Matthews, 2011). Cortisol and testosterone in humans have been positively associated in amniotic fluid (Sarkar, Bergman, Fisk, O'Connor, & Glover, 2007), and an overlap has been reported between corticotropin-releasing hormone and estriol in maternal serum in pregnancy (Smith et al., 2009).

There are few discrete examples in the human research that PMD may alter or program sex steroids in the child, with implications for brain and behavior. Two available examples include anogenital distance, a biomarker of prenatal androgen exposure (Barrett et al., 2013) and play behavior (Barrett, Redmon, Wang, Sparks, & Swan, 2014). These results raise the important possibility that sex steroids may carry some of the effects of PMD on child outcomes and underscore the inevitable confound between HPA axis and glucocorticoids and the hypothalamic pituitary gonadal (HPG) axis and sex steroid pathways.

The substantial research base assessing stress and laterality in animal and human models (e.g., Ocklenburg, Korte, Peterburs, Wolf, & Gunturkun, 2016) now needs translation and application in research on maternal prenatal stress and child health. Such research would help clarify how and if the PMD and GGH are compatible and would offer leverage for testing novel hypotheses, such as the possibility that the lateralized effects of prenatal maternal stress on the developing child's brain explains stress reactivity and behavioral health outcomes in the child.

We also note some important methodological considerations for further work. The inadequacy of the murine model because the mouse lacks adrenal androgens and placental estrogens is most pressing for translational research (Thibeault, Deroy, Vaillancourt, & Sanderson, 2014; van Weerden, Bierings, van Steenbrugge, de Jong, & Schroder, 1992). A more relevant model is the guinea pig, in which findings indicate that females exposed to prenatal stressors exhibit masculinized sexual and social behavior in adulthood and male-typical patterns of estrogen and androgen receptors in sexually dimorphic brain nuclei (Kaiser, Kruijver, Swaab, & Sachser, 2003; Kaiser & Sachser, 1998). Animal research is necessary and helpful in this area because of the other major obstacle for human research—the difficulty in sampling (directly or indirectly) relevant molecules at the key points in development.

We are currently investigating the novel hypothesis that PMD induces upregulation of fetal adrenal androgen production, leading to downstream changes in feto-placental sex steroid pathways, with implications for social behavior, neurodevelopment, and somatic health. Examining the extent to which anxiety alters fetal adrenal androgen pathways by measuring maternal estriol E3 is particularly important in the model, given the limits of measurement in humans (see Figure 5.1).

That pathway is as follows: CRH stimulates dehydroepiandrosterone sulfate (DHEAS) production by the fetal adrenal; DHEAS in the fetal liver is then converted to 16αOH-DHEAS, the precursor for placental E3 production. Placental E3, which reflects fetal adrenal hormone production, accounts for nearly all circulating maternal E3 (Smith, Paul, Maiti, Tolosa, & Madsen, 2012) and, therefore, gives unique insight into fetal adrenal function. The model has particular advantages conceptually; methodologically, other hormones in maternal circulation typically derive from

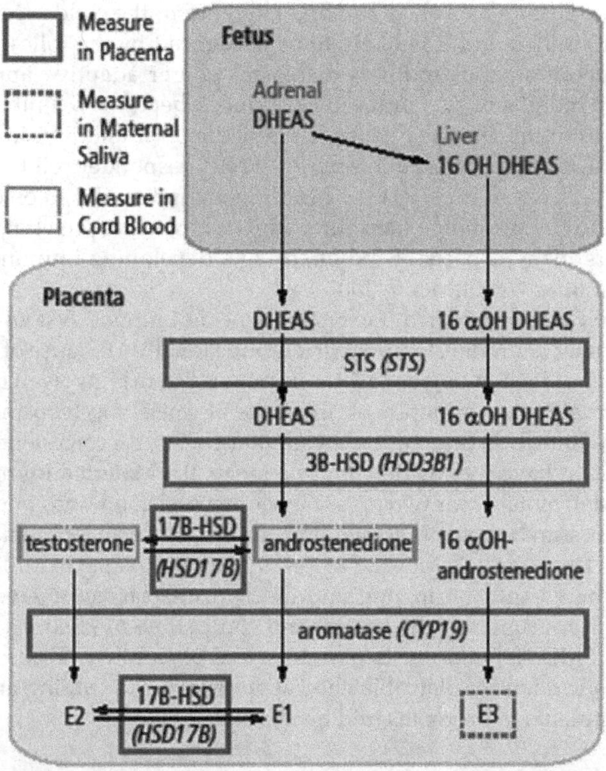

Figure 5.1 The Figure displays the hypothesized biological pathway by which the measurement of estriol (E3) in maternal samples may offer an index of fetal androgen activity. DHEAS = dehydroepiandrosterone sulfate; 16(α)OH DHEAS= 16 (alpha) hydroxy DHEAS; STS = steroid sulfatase; HSD hydroxysteroid dehydrogenase (the Figure includes several forms); E = estrogens measured in maternal samples: E1: estrone; E2: estradiol; E3: estriol. A full-color version of this figure is available online at http://www.brookespublishing.com/galaburda/materials.

multiple compartments (i.e., mother, placenta, fetus) in unknown ratios, making them unreliable indices of the fetal hormonal milieu and of limited explanatory value in the human model.

PRENATAL MATERNAL DISTRESS AND CHILD NEUROENDOCRINE AND IMMUNOLOGICAL OUTCOMES

The final section explores the provocative component of the GGH that suggests factors that may modify "cerebral dominance also influence the development of many other systems, e.g., the organs involved in the immune response." (Geschwind & Galaburda, 1985a, p. 431). PMD may influence the immune system as well as cerebral laterality in the child. Work in this area is still limited, but it is likely to be influenced by the following findings. Infants of anxious mothers displayed poorer adaptive immune response at 6 months of age, indexed by reduced hepatitis B antibody titers following immunization, as well as cell-mediated immunity, indexed by increased IL-4 and decreased interferon(IFN)-γ responder cell frequencies to antigen (O'Connor et al., 2013). These human findings build on considerable evidence in the animal literature, which shows that prenatal maternal stress alters many aspects of the offspring's developing immune system (e.g., Coe, Lubach, & Shirtcliff, 2007).

As the GGH anticipated, factors influencing immune system development may alter neurodevelopment (including laterality) because of the now-appreciated understanding of how the immune system is involved in healthy and abnormal brain development. Evidence of causal mechanisms is obviously more limited in human research, but associations consistent with the general notion have been widely reported (e.g., the Maternal Immune Activation model), notably for two neurodevelopmental conditions of particular interest to research on cerebral laterality, autism, and schizophrenia (Brown et al., 2004; Depino, 2013; Meyer, Feldon, & Yee, 2009; Patterson, 2011).

The shared interest in immune mechanisms on neurodevelopment is an important aspect of the shared and compatible predictions from the GGH and PMD hypotheses. Exploratory and elaborative research in this area may also identify plausible clinical application for timing and monitoring treatments to promote child health.

CONCLUSION

A working thesis in this chapter is that important parallels exist between the GGH and developing research on PMD. Evidence is sizable, and different kinds of evidence have been provided. A stronger impression, however, is that the overlap in research literatures is surreptitious and incomplete because many conditions of particular relevance to the GGH have not yet been studied in relation to prenatal maternal stress, and the existing literature on prenatal distress has not yet been viewed through the GGH filter. Therefore, we encourage further study of the compatibilities and contrasts in these largely parallel literatures to promote understanding of the origins of developmental disorders and their prevention and treatment.

REFERENCES

Ader, R., & Plaut, S. M. (1968). Effects of prenatal maternal handling and differential housing on offspring emotionality, plasma corticosterone levels, and susceptibility to gastric erosions. *Psychosomatic Medicine, 30*(3), 277–286.

Auyeung, B., Ahluwalia, J., Thomson, L., Taylor, K., Hackett, G., O'Donnell, K. J., & Baron-Cohen, S. (2012). Prenatal versus postnatal sex steroid hormone effects on autistic traits in children at 18 to 24 months of age. *Molecular Autism, 3*(1), 17. doi:10.1186/2040-2392-3-17

Barker, D. J. (1993). Fetal origins of coronary heart disease. *British Heart Journal, 69*(3), 195–196.

Barrett, E. S., Parlett, L. E., Sathyanarayana, S., Liu, F., Redmon, J. B., Wang, C., & Swan, S. H. (2013). Prenatal exposure to stressful life events is associated with masculinized anogenital distance (AGD) in female infants. *Physiology and Behavior, 114-115,* 14–20. doi:10.1016/j.physbeh.2013.03.004

Barrett, E. S., Redmon, J. B., Wang, C., Sparks, A., & Swan, S. H. (2014). Exposure to prenatal life events stress is associated with masculinized play behavior in girls. *Neurotoxicology, 41C,* 20–27. doi:10.1016/j.neuro.2013.12.011

Bergman, K., Sarkar, P., Glover, V., & O'Connor, T. G. (2010). Maternal prenatal cortisol and infant cognitive development: Moderation by infant-mother attachment. *Biological Psychiatry, 67*(11), 1026–1032. doi:S0006-3223(10)00013-2 [pii]10.1016/j.biopsych.2010.01.002

Bilbo, S. D., & Schwarz, J. M. (2009). Early-life programming of later-life brain and behavior: A critical role for the immune system. *Frontiers of Behavioral Neuroscience, 3,* 14. doi:10.3389/neuro.08.014.2009

Brown, A. S., Hooton, J., Schaefer, C. A., Zhang, H., Petkova, E., Babulas, V.,…Susser, E. S. (2004). Elevated maternal interleukin-8 levels and risk of schizophrenia in adult offspring. *American Journal of Psychiatry, 161*(5), 889–895.

Buss, C., Davis, E. P., Hobel, C. J., & Sandman, C. A. (2011). Maternal pregnancy-specific anxiety is associated with child executive function at 6-9 years age. *Stress, 14*(6), 665–676. doi:10.3109/10253890.2011.623250

Coe, C. L., Lubach, G. R., & Karaszewski, J. W. (1999). Prenatal stress and immune recognition of self and nonself in the primate neonate. *Biology of the Neonate, 76*(5), 301–310. doi:bon76301 [pii]

Coe, C. L., Lubach, G. R., & Schneider, M. L. (2002). Prenatal disturbance alters the size of the corpus callosum in young monkeys. *Developmental Psychobiology, 41*(2), 178–185. doi:10.1002/dev.10063

Coe, C. L., Lubach, G. R., & Shirtcliff, E. A. (2007). Maternal stress during pregnancy predisposes for iron deficiency in infant monkeys impacting innate immunity. *Pediatric Research, 61*(5 Pt 1), 520–524. doi:10.1203/pdr.0b013e318045be53

Cohen-Bendahan, C. C., Buitelaar, J. K., van Goozen, S. H., & Cohen-Kettenis, P. T. (2004). Prenatal exposure to testosterone and functional cerebral lateralization: A study in same-sex and opposite-sex twin girls. *Psychoneuroendocrinology, 29*(7), 911–916. doi:10.1016/j.psyneuen.2003.07.001

Darnaudery, M., & Maccari, S. (2008). Epigenetic programming of the stress response in male and female rats by prenatal restraint stress. *Brain Research Review, 57*(2), 571–585. doi:S0165-0173(07)00264-0 [pii] 10.1016/j.brainresrev.2007.11.004

Depino, A. M. (2013). Peripheral and central inflammation in autism spectrum disorders. *Molecular and Cellular Neuroscience, 53,* 69–76. doi:10.1016/j.mcn.2012.10.003

Entringer, S., Buss, C., Andersen, J., Chicz-Demet, A., & Wadhwa, P. D. (2011). Ecological momentary assessment of maternal cortisol profiles over a multiple-day period predicts the length of human gestation. *Psychosomatic Medicine, 73*(6), 469–474. doi:PSY.0b013e31821fbf9a [pii] 10.1097/PSY.0b013e31821fbf9a

Entringer, S., Buss, C., & Wadhwa, P. D. (2010). Prenatal stress and developmental programming of human health and disease risk: Concepts and integration of empirical findings. *Current Opinion in Endocrinology, Diabetes, and Obesity, 17*(6), 507–516. doi:10.1097/MED.0b013e3283405921

Entringer, S., Kumsta, R., Nelson, E. L., Hellhammer, D. H., Wadhwa, P. D., & Wust, S. (2008). Influence of prenatal psychosocial stress on cytokine production in adult women. *Developmental Psychobiology, 50*(6), 579–587. doi:10.1002/dev.20316

Geschwind, N., & Galaburda, A. M. (1985a). Cerebral lateralization. Biological mechanisms, associations, and pathology: I. A hypothesis and a program for research. *Archives of Neurology, 42*(5), 428–459.

Geschwind, N., & Galaburda, A. M. (1985b). Cerebral lateralization. Biological mechanisms, associations, and pathology: II. A hypothesis and a program for research. *Archives of Neurology, 42*(6), 521–552.

Geschwind, N., & Galaburda, A. M. (1985c). Cerebral lateralization. Biological mechanisms, associations, and pathology: III. A hypothesis and a program for research. *Archives of Neurology, 42*(7), 634–654.

Glover, V., O'Connor, T. G., Heron, J., & Golding, J. (2004). Antenatal maternal anxiety is linked with atypical handedness in the child. *Early Human Development, 79*(2), 107–118. doi:10.1016/j.earlhumdev.2004.04.012 S0378-3782(04)00072-6 [pii]

Glover, V., O'Connor, T. G., & O'Donnell, K. (2010). Prenatal stress and the programming of the HPA axis. *Neuroscience and Biobehavior Reviews, 35*(1), 17–22. doi:S0149-7634(09)00174-2 [pii] 10.1016/j.neubiorev.2009.11.008

Gluckman, P., & Hanson, M. (2005). *The fetal matrix: Evolution, development, and disease.* Cambridge, UK: Cambridge University Press.

Godfrey, K. M., Gluckman, P. D., & Hanson, M. A. (2010). Developmental origins of metabolic disease: Life course and intergenerational perspectives. *Trends in Endocrinology and Metabolism, 21*(4), 199–205. doi:S1043-2760(09)00219-7 [pii] 10.1016/j.tem.2009.12.008

Gutteling, B. M., de Weerth, C., & Buitelaar, J. K. (2007). Prenatal stress and mixed-handedness. *Pediatric Research, 62*(5), 586–590. doi:10.1203/PDR.0b013e3181558678

Hale, T. S., Loo, S. K., Zaidel, E., Hanada, G., Macion, J., & Smalley, S. L. (2009). Rethinking a right hemisphere deficit in ADHD. *Journal of Attention Disorders, 13*(1), 3–17. doi:10.1177/1087054708323005

Hales, C. N., & Barker, D. J. (1992). Type 2 (non-insulin-dependent) diabetes mellitus: The thrifty phenotype hypothesis. *Diabetologia, 35*(7), 595–601.

Hanson, M., & Gluckman, P. (2011). Developmental origins of noncommunicable disease: Population and public health implications. *American Journal of Clinical Nutrition, 94*(6), 17545s–17585s. doi:ajcn.110.001206 [pii] 10.3945/ajcn.110.001206

Henry, C., Kabbaj, M., Simon, H., Le Moal, M., & Maccari, S. (1994). Prenatal stress increases the hypothalamo-pituitary-adrenal axis response in young and adult rats. *Journal of Neuroendocrinology, 6*(3), 341–345.

Hines, M., Fane, B. A., Pasterski, V. L., Mathews, G. A., Conway, G. S., & Brook, C. (2003). Spatial abilities following prenatal androgen abnormality: Targeting and mental rotations performance in individuals with congenital adrenal hyperplasia. *Psychoneuroendocrinology, 28*(8), 1010–1026.

Jones, A., Osmond, C., Godfrey, K. M., & Phillips, D. I. (2011). Evidence for developmental programming of cerebral laterality in humans. *PLoS One, 6*(2), e17071. doi:10.1371/journal.pone.0017071

Kaiser, S., Kruijver, F. P., Swaab, D. F., & Sachser, N. (2003). Early social stress in female guinea pigs induces a masculinization of adult behavior and corresponding changes in brain and neuroendocrine function. *Behavioural Brain Research, 144*(1-2), 199–210.

Kaiser, S., & Sachser, N. (1998). The social environment during pregnancy and lactation affects the female offsprings' endocrine status and behaviour in guinea pigs. *Physiology and Behavior, 63*(3), 361–366.

Kapoor, A., & Matthews, S. G. (2011). Testosterone is involved in mediating the effects of prenatal stress in male guinea pig offspring. *Journal of Physiology, 589*(Pt 3), 755–766. doi:jphysiol.2010.200543 [pii] 10.1113/jphysiol.2010.200543

Khashan, A. S., Abel, K. M., McNamee, R., Pedersen, M. G., Webb, R. T., Baker, P. N.,… Mortensen, P. B. (2008). Higher risk of offspring schizophrenia following antenatal maternal exposure to severe adverse life events. *Archives of General Psychiatry, 65*(2), 146–152. doi:10.1001/archgenpsychiatry.2007.20

King, S., Mancini-Marie, A., Brunet, A., Walker, E., Meaney, M. J., & Laplante, D. P. (2009). Prenatal maternal stress from a natural disaster predicts dermatoglyphic asymmetry in humans. *Development and Psychopathology, 21*(2), 343–353. doi:10.1017/S0954579409000364

Kinney, D. K., Munir, K. M., Crowley, D. J., & Miller, A. M. (2008). Prenatal stress and risk for autism. *Neuroscience Biobehavior Review, 32*(8), 1519–1532. doi:S0149-7634(08)00098-5 [pii] 10.1016/j.neubiorev.2008.06.004

Kung, K. T., Spencer, D., Pasterski, V., Neufeld, S., Glover, V., O'Connor, T. G.,…Hines, M. (2016). No relationship between prenatal androgen exposure and autistic traits: Convergent evidence from studies of children with congenital adrenal hyperplasia and of amniotic testosterone concentrations in typically developing children. *Journal of Child Psychology and Psychiatry 57*(12), 1455–1462. doi:10.1111/jcpp.12602

Laplante, D. P., Barr, R. G., Brunet, A., Galbaud du Fort, G., Meaney, M. L., Saucier, J. F.,… King, S. (2004). Stress during pregnancy affects general intellectual and language functioning in human toddlers. *Pediatric Research, 56*(3), 400–410. doi:10.1203/01.PDR.0000136281.34035.44

Li, J., Olsen, J., Vestergaard, M., & Obel, C. (2010). Attention-deficit/hyperactivity disorder in the offspring following prenatal maternal bereavement: A nationwide follow-up study in Denmark. *European Child and Adolescent Psychiatry, 19*(10), 747–753. doi:10.1007/s00787-010-0113-9

Li, J., Vestergaard, M., Obel, C., Christensen, J., Precht, D. H., Lu, M., & Olsen, J. (2009). A nationwide study on the risk of autism after prenatal stress exposure to maternal bereavement. *Pediatrics, 123*(4), 1102–1107. doi:123/4/1102 [pii] 10.1542/peds.2008-1734

Lindell, A. K., & Hudry, K. (2013). Atypicalities in cortical structure, handedness, and functional lateralization for language in autism spectrum disorders. *Neuropsychology Review, 23*(3), 257–270. doi:10.1007/s11065-013-9234-5

Lobel, M., Dunkel-Schetter, C., & Scrimshaw, S. C. (1992). Prenatal maternal stress and prematurity: A prospective study of socioeconomically disadvantaged women. *Health Psychology, 11*(1), 32–40.

MacLusky, N. J., & Naftolin, F. (1981). Sexual differentiation of the central nervous system. *Science, 211*(4488), 1294–1302.

Matthews, S. G. (2000). Antenatal glucocorticoids and programming of the developing CNS. *Pediatric Research, 47*(3), 291–300.

Meyer, U., Feldon, J., & Yee, B. K. (2009). A review of the fetal brain cytokine imbalance hypothesis of schizophrenia. *Schizophrenia Bulletin, 35*(5), 959–972. doi:10.1093/schbul/sbn022

Nasreen, H. E., Kabir, Z. N., Forsell, Y., & Edhborg, M. (2011). Prevalence and associated factors of depressive and anxiety symptoms during pregnancy: A population based study in rural Bangladesh. *BMC Womens Health, 11*, 22. doi:1472-6874-11-22 [pii] 10.1186/1472-6874-11-22

Obel, C., Hedegaard, M., Henriksen, T. B., Secher, N. J., & Olsen, J. (2003). Psychological factors in pregnancy and mixed-handedness in the offspring. *Developmental Medicine and Child Neurology, 45*(8), 557–561.

Ocklenburg, S., Korte, S. M., Peterburs, J., Wolf, O. T., & Gunturkun, O. (2016). Stress and laterality: The comparative perspective. *Physiology and Behavior, 164*(Pt A), 321–329. doi: 10.1016/j.physbeh.2016.06.020

O'Connor, T. G., & Barrett, E. S. (2014). Mechanisms of prenatal programing: Identifying and distinguishing the impact of steroid hormones. *Frontiers in Endocrinology, 5*, 52. doi:10.3389/fendo.2014.00052

O'Connor, T. G., Monk, C., & Fitelson, E. M. (2014). Practitioner review: Maternal mood in pregnancy and child development—implications for child psychology and psychiatry. *Journal of Child Psychology and Psychiatry and Allied Disciplines, 55*(2), 99–111. doi:10.1111/jcpp.12153

O'Connor, T. G., Winter, M. A., Hunn, J., Carnahan, J., Pressman, E. K., Glover, V.,… Caserta, M. T. (2013). Prenatal maternal anxiety predicts reduced adaptive immunity in infants. *Brain, Behavior, and Immunity, 32*, 21–28. doi:10.1016/j.bbi.2013.02.002

O'Donnell, K. J., Bugge Jensen, A., Freeman, L., Khalife, N., O'Connor, T. G., & Glover, V. (2012). Maternal prenatal anxiety and downregulation of placental 11beta-HSD2. *Psychoneuroendocrinology, 37*(6), 818–826. doi:10.1016/j.psyneuen.2011.09.014

O'Donnell, K. J., Glover, V., Barker, E. D., & O'Connor, T. G. (2014). The persisting effect of maternal mood in pregnancy on childhood psychopathology. *Development and Psychopathology, 26*(2), 393–403. doi:10.1017/S0954579414000029

Owen, D., Andrews, M. H., & Matthews, S. G. (2005). Maternal adversity, glucocorticoids and programming of neuroendocrine function and behaviour. *Neuroscience and Biobehavioral Review, 29*(2), 209–226. doi:S0149-7634(04)00119-8 [pii] 10.1016/j.neubiorev.2004.10.004

Patterson, P. H. (2011). Maternal infection and immune involvement in autism. *Trends in Molecular Medicine, 17*(7), 389–394. doi:10.1016/j.molmed.2011.03.001

Rifkin-Graboi, A., Meaney, M. J., Chen, H., Bai, J., Hameed, W. B., Tint, M. T.,…Qiu, A. (2015). Antenatal maternal anxiety predicts variations in neural structures implicated in anxiety disorders in newborns. *Journal of the American Academy of Child and Adolescent Psychiatry, 54*(4), 313–321. e312. doi:10.1016/j.jaac.2015.01.013

Ross, L. E., & McLean, L. M. (2006). Anxiety disorders during pregnancy and the postpartum period: A systematic review. *Journal of Clinical Psychiatry, 67*(8), 1285–1298.

Sarkar, P., Bergman, K., Fisk, N. M., O'Connor, T. G., & Glover, V. (2007). Amniotic fluid testosterone: Relationship with cortisol and gestational age. *Clinical Endocrinology, 67*(5), 743–747. doi:10.1111/j.1365-2265.2007.02955.x

Sarkar, S., Craig, M. C., Dell'acqua, F., O'Connor, T. G., Catani, M., Deeley, Q.,…Murphy, D. G. (2014). Prenatal stress and limbic-prefrontal white matter microstructure in children aged 6-9 years: A preliminary diffusion tensor imaging study. *World Journal of Biological Psychiatry, 15*(4), 346–352. doi:10.3109/15622975.2014.903336

Seckl, J. R., & Meaney, M. J. (2004). Glucocorticoid programming. *Annals of the New York Academy of Sciences, 1032*, 63–84. doi:10.1196/annals.1314.006

Smith, R., Paul, J., Maiti, K., Tolosa, J., & Madsen, G. (2012). Recent advances in understanding the endocrinology of human birth. *Trends in Endocrinology and Metabolism, 23*(10), 516–523. doi:10.1016/j.tem.2012.05.002

Smith, R., Smith, J. I., Shen, X., Engel, P. J., Bowman, M. E., McGrath, S. A.,…Smith, D. W. (2009). Patterns of plasma corticotropin-releasing hormone, progesterone, estradiol, and estriol change and the onset of human labor. *Journal of Clinical Endocrinology and Metabolism, 94*(6), 2066–2074. doi:10.1210/jc.2008-2257

Swan, S. H., Liu, F., Hines, M., Kruse, R. L., Wang, C., Redmon, J. B.,…Weiss, B. (2010). Prenatal phthalate exposure and reduced masculine play in boys. *International Journal of Andrology, 33*(2), 259–269. doi:10.1111/j.1365-2605.2009.01019.x

Talge, N. M., Neal, C., & Glover, V. (2007). Antenatal maternal stress and long-term effects on child neurodevelopment: How and why? *Journal of Child Psychology and Psychiatry, 48*(3-4), 245–261. doi:JCPP1714 [pii]10.1111/j.1469-7610.2006.01714.x

Thibeault, A. A., Deroy, K., Vaillancourt, C., & Sanderson, J. T. (2014). A unique co-culture model for fundamental and applied studies of human fetoplacental steroidogenesis and interference by environmental chemicals. *Environmental Health Perspectives, 122*(4), 371–377. doi:10.1289/ehp.1307518

van Batenburg-Eddes, T., de Groot, L., Huizink, A. C., Steegers, E. A., Hofman, A., Jaddoe, V. W.,…Tiemeier, H. (2009). Maternal symptoms of anxiety during pregnancy affect infant neuromotor development: The generation R study. *Developmental Neuropsychology, 34*(4), 476–493. doi:10.1080/87565640902964508

van den Bergh, B. R., Mennes, M., Stevens, V., van der Meere, J., Borger, N., Stiers, P.,… Lagae, L. (2006). ADHD deficit as measured in adolescent boys with a continuous performance task is related to antenatal maternal anxiety. *Pediatric Research, 59*(1), 78–82. doi:10.1203/01.pdr.0000191143.75673.52

van Weerden, W. M., Bierings, H. G., van Steenbrugge, G. J., de Jong, F. H., & Schroder, F. H. (1992). Adrenal glands of mouse and rat do not synthesize androgens. *Life Sciences, 50*(12), 857–861.

Wadhwa, P. D., Dunkel-Schetter, C., Chicz-DeMet, A., Porto, M., & Sandman, C. A. (1996). Prenatal psychosocial factors and the neuroendocrine axis in human pregnancy. *Psychosomatic Medicine, 58*(5), 432–446.

CHAPTER 6

Pubertal Effects of Androgens on Brain Development and Lateralization

Tuong-Vi Nguyen

SUMMARY

The biological processes underpinning brain development during middle childhood (6–11 years old) are particularly important to elucidate, given the emergence of new cognitive skills during this period, including reading and writing. Such development coincides with an increase in specific hormones called *androgens,* including dehydroepiandrosterone (DHEA). Only humans and the great apes have been demonstrated to exhibit this rise in DHEA, suggesting it may play an important evolutionary role. In fact, DHEA represents the most abundant steroid hormone in circulation across the life span. Evidence at the molecular level has also confirmed the importance of DHEA in neuroprotection.

Although DHEA represents the primary hormone in circulation during middle childhood, production of testosterone may also play a role during this developmental period. Higher testosterone levels tend to be associated with neurotoxic effects. Testosterone is thought to play an important role in determining sex differences in brain structure and behavior. In fact, the Geschwind-Galaburda Hypothesis (GGH) proposed that testosterone plays an important role in hemispheric lateralization in which the left hemisphere of males undergoes more restricted growth than the right hemisphere prenatally (Geschwind & Galaburda, 1985a, b, c). DHEA and testosterone appear to have competing and antagonistic effects on the same final common pathways, which suggests that androgens may interact in shaping brain development. This interaction likely varies across the age span, depending on the respective levels of each hormone. This chapter first discusses the role of androgens in trajectories of cortical maturation. Then it reviews evidence linking androgens, cortico-limbic maturation, and developmental changes in cognition and behavior. Finally, it discusses the developmental role of androgens in the central nervous system (CNS) in the context of the GGH, and the chapter ends with a review of future avenues of research.

INTRODUCTION

The molecular and endocrine processes underpinning brain development during middle childhood (6–11 years old) are particularly important to elucidate, given the emergence of new developmental skills during this period, including reading and writing. This neurobehavioral development coincides with adrenarche, a key pubertal event that precedes and then parallels gonadarche (i.e., ovarian and testicular maturation). A significant rise in the adrenal production of androgens occurs during adrenarche, in particular, DHEA and its main metabolites, which increase significantly from ages 3 to 13, remain at a steady level until the third decade of life, and decrease thereafter, coinciding with synaptic loss/cortical thinning. Interestingly, humans and the great apes are the only species demonstrated to exhibit adrenarche, suggesting that DHEA may play an important evolutionary role (Remer, Boye, Hartmann, & Wudy, 2005).

In fact, DHEA and its derivative dehydroepiandrosterone sulfate (DHEAS) represent the most abundant steroid hormones in circulation across the life span. Evidence at the molecular level has also confirmed the important role of DHEA in axonal/dendritic growth and neuroprotection (Maninger, Wolkowitz, Reus, Epel, & Mellon, 2009). In contrast, DHEAS was reported to increase apoptosis in neural precursors under specific circumstances.

Although DHEA and DHEAS represent the primary steroid hormones in circulation during middle childhood, adrenal production of testosterone may also play a role during this developmental period. Prior findings from our group showed that higher testosterone levels tend to be associated with decreased cortical thickness (CTh; see Chapter 8) during adolescence (Nguyen, McCracken, Ducharme, Botteron et al., 2013), consistent with in vitro evidence that high testosterone levels may exert proapoptotic effects. Yet, in other settings where lower testosterone levels predominate, it may be associated with protection from programmed cell death and increased CTh (Nguyen, McCracken, Ducharme, Botteron et al., 2013; Vasconsuelo, Pronsato, Ronda, Boland, & Milanesi, 2011).

Testosterone is thought to play an important role in sexual differentiation of the CNS as shown by evidence of permanent and irreversible sex differences related to testosterone in cortical and subcortical brain structures. GGH also proposes that testosterone plays an important role in hemispheric lateralization in which the left hemisphere of males undergoes more restricted growth than the right hemisphere in utero (Geschwind & Galaburda, 1985a, b, c).

DHEA and testosterone appear to have competing and antagonistic effects on the same final common pathways, such as the modulation of intracellular calcium levels. This suggests that androgens may interact in shaping brain development, and this interaction likely varies across the age span, depending on the respective levels of each steroid hormone.

In fact, there is accumulating evidence that androgens carry out important developmental functions in the context of 1) cortical maturation, 2) cortico-limbic structural networks, and 3) cognitive control and behavioral regulation during middle childhood and adolescence. As detailed next, the role of androgens in trajectories of cortical maturation are discussed. Then, evidence linking androgens, cortico-limbic maturation, and developmental changes in cognition and behavior are reviewed. Finally, the developmental role of androgens in the CNS in the context of the GGH is discussed, and the chapter ends with a review of future avenues of research.

DHEA-related Cortical Maturation During Development

DHEA was found to be associated with CTh increases from ages 4 to 13 years in a large longitudinal sample of children and adolescents who were 4–22 years of age (Nguyen, McCracken, Ducharme, Cropp et al., 2013), supporting the concept that the effects of DHEA are time and dose dependent, with certain ages and doses associated with different neurotrophic effects. These effects were not lateralized to a particular hemisphere and were independent from sex. Pronounced regional effects were found, with DHEA-related increases in CTh occurring in specific brain areas (Nguyen, McCracken, Ducharme, Cropp et al., 2013). These DHEA-responsive brain areas may represent sites of neurotrophic changes during significant developmental processes of middle childhood. For example, the association of DHEA with CTh of the right tempoparietal junction in children ages 7–12 is consistent with the development of mentalizing and differentiation of this region between 7 and 11 years old (Saxe, Whitfield-Gabrieli, Scholz, & Pelphrey, 2009). DHEA is also associated with CTh of the left dorsolateral prefrontal cortex between ages 4 and 8, paralleling reports of increased activation in this region associated with improved language/behavior control across the same age range (Blakemore, 2008; Szaflarski et al., 2006). DHEA-related maturation of the right premotor region between ages 5 and 11 partially overlaps with previous reports of developmental changes in spatial working memory in children ages 8–11 (Nelson et al., 2000). Finally, DHEA was associated with changes in CTh of the right entorhinal cortex between ages 4 and 13. This is consistent with the extended gray matter preservation in this region from childhood to the third decade of life, a pattern hypothesized to be related to the ongoing development of emotional processing/memory (Grieve, Korgaonkar, Clark, & Williams, 2011).

Testosterone-related Cortical Maturation During Development

In contrast to DHEA, testosterone showed sex- and hemisphere-specific effects on CTh (Nguyen, McCracken, Ducharme, Botteron et al., 2013). Indeed, testosterone was found to be negatively associated with CTh of the left hemisphere in postpubertal boys between the ages of 14 and 22 years, more specifically the posterior cingulate cortex (PCC), precuneus, and dorsolateral prefrontal cortex, and later the anterior cingulate cortex (ACC). These negative associations are consistent with in vitro evidence that high testosterone levels may exert proapoptotic effects. In contrast, early prepubertal girls 4–7 years old showed a positive association between CTh and testosterone levels in the right hemisphere, involving the somatosensory cortex, a relationship that was reversed in late postpubertal females 21–22 years old with a negative association in the same area (Nguyen, McCracken, Ducharme, Botteron et al., 2013), consistent with the observed neuroprotective influence of testosterone in developing female animal models.

The finding that peripubertal testosterone levels affected CTh in the left hemisphere more than the right hemisphere in boys, whereas the opposite pattern was present in girls (Nguyen, McCracken, Ducharme, Botteron et al., 2013), persisted even when controlling for handedness, supporting the view that manual dominance likely follows a different process from lateralization of cognitive functions such as verbal or visuospatial dominance (Geschwind & Galaburda, 1985a, b, c). This relationship between testosterone and lateralization of CTh maturation patterns was also shown to be age specific and to evolve throughout puberty (Nguyen, McCracken, Ducharme, Botteron et al., 2013). These results suggest that lateralization follows a different pattern in males and females as well as represents a more dynamic process than previously thought. This is partly supported by previous investigations into interhemispheric inhibition of the nondominant hemisphere (by the dominant hemisphere) and functional organization within hemispheres, showing that these functional cerebral asymmetries are sex specific and sex-hormone dependent, being relatively stable in men and fluctuating in women across the menstrual cycle.

Interactive Effects of DHEA and Testosterone During Development

Adding to the complexity of individual hormonal effects, high DHEA levels were shown to mitigate the negative relationship between testosterone levels and CTh in a high-testosterone context, whereas the opposite relationship was observed in a low-testosterone context, such that testosterone mitigated the negative relationship between DHEA and CTh (Nguyen, McCracken, Ducharme, Cropp et al., 2013) (see Figure 6.1). This complex interaction is consistent with previous reports of both proliferative and antiproliferative changes observed with DHEA and testosterone in vitro. These paradoxical effects may rely on two competing and time-dependent androgen pathways—a slower genomic, nuclear receptor pathway that relies mostly on changes in DNA transcription and a faster, nongenomic membrane receptor pathway relying mainly on second-messenger signaling systems. The fast-acting, nongenomic pathway appears to be most active when high hormonal levels predominate, whereas the slower genomic pathway may be more active when lower hormonal levels are present.

These interactive effects of testosterone and DHEA were found to affect CTh of the orbitofrontal cortex (OFC), ACC, PCC, and occipital pole

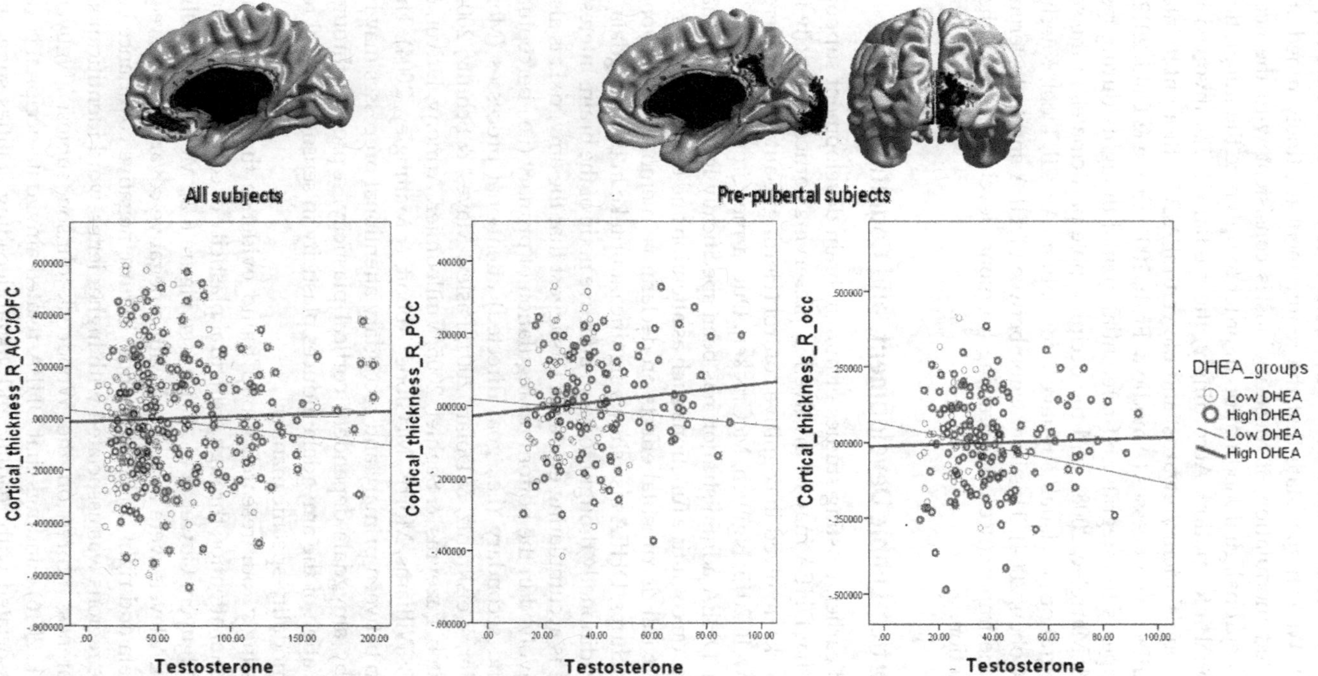

Figure 6.1. The brain figures show regions with a significant DHEA by testosterone interaction on cortical thickness (CTh): the right anterior cingulate cortex and orbitofrontal cortex, the right posterior cingulate cortex, and the right occipital pole. Higher DHEA levels mitigate the negative relationship between high testosterone levels and CTh, whereas the opposite relationship is present in a low-testosterone context, such that testosterone mitigates the negative relationship between DHEA and CTh. (From Nguyen, McCracken, Ducharme, Cropp et al. [2013]. *Journal of Neuroscience 33*[26], 10840–10848. doi: 10.1523/JNEUROSCI.5747-12.2013; republished with permission of the authors permission conveyed through Copyright Clearance Center.)

(OP) (Nguyen, Gower et al., 2016; Nguyen, McCracken, Ducharme, Cropp et al., 2013). We can speculate that the requirement for finely tuned androgen-related modulation of these regions is consistent with the need for specific neuronal/glial modulation in order to optimize the integration of multimodal information. Accordingly, the medial OFC is involved in monitoring punishments/rewards, and the ventral ACC is thought to integrate socioaffective processes (Amodio & Frith, 2006; Crone & Dahl, 2012); the PCC appears to integrate self-referential mental thoughts during rest (Fransson & Marrelec, 2008); and the occipital pole participates in the semantic integration of visual/language information (Amedi, Floel, Knecht, Zohary, & Cohen, 2004). This interaction between DHEA and testosterone highlights the importance of appraising hormonal levels together rather than in isolation.

DHEA, Cortico-Limbic Development, and Cognition

As detailed earlier, existing studies of human brain development support the notion that DHEA may be involved in preserving cortical plasticity in brain networks involved in cognitive control (Herting, Gautam, Spielberg, Dahl, & Sowell, 2015; Nguyen, McCracken, Ducharme, Cropp et al., 2013). In addition, DHEA administration has been specifically linked to optimal performance on cognitive functions such as attention and executive control, with paradoxical detrimental effects on declarative memory. Yet, beyond cortical structures, DHEA also shows significant uptake in the amygdala, a structure traditionally thought to be an affective node of the brain. Interestingly, there is accumulating evidence to suggest that the amygdala is also centrally involved in the bottom-up regulation of primary (i.e., perceptual salience) and secondary (i.e., goal directed) attentional processes (Morris, DeGelder, Weiskrantz, & Dolan, 2001; Pasley, Mayes, & Schultz, 2004; Pegna, Khateb, Lazeyras, & Seghier, 2005; Vuilleumier, Armony, Driver, & Dolan, 2003; Williams, Morris, McGlone, Abbott, & Mattingley, 2004). The relationship between primary and secondary attentional processes may be facilitated by amygdala-dependent cortical plasticity, a process through which activation of the amygdala induces plasticity in sensory cortical regions in a modality-specific manner.

Interestingly, our research group found evidence that DHEA may regulate visual attention through structural plasticity of cortico-amygdalar networks (Nguyen, Gower et al., 2016) (see Figure 6.2). More specifically, higher DHEA levels were associated with a negative covariance between left amygdala and right occipital regions; in turn, negative covariance between these regions was associated with higher letter–word identification/visual awareness scores on the Woodcock-Johnson subtest (Nguyen, Gower et al., 2016). Findings that DHEA-related amygdala-occipital covariance is associated with reading and visual decoding abilities suggest a potential role for DHEA in the filtering of salient visual stimuli and visual

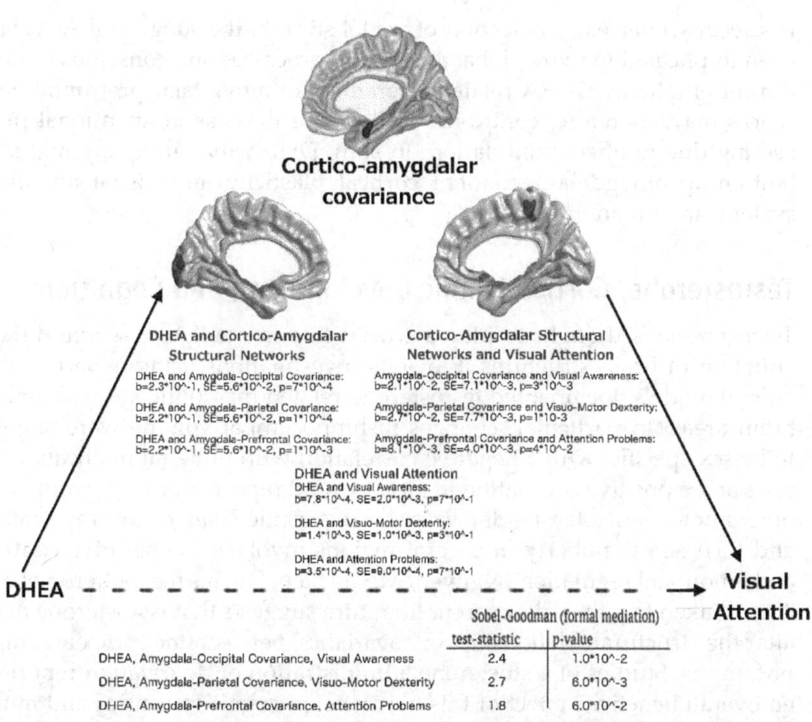

Figure 6.2. This figure displays the relationship between DHEA levels, cortico-amygdalar covariance, and visual attention. The brain figures show the regions where the relationship between cortical thickness and left and right amygdalar volume significantly varies according to DHEA levels. Beta coefficients, standard errors, and *p*-values are displayed in the center of the figure. (From Nguyen, Gower, Albaugh, Botteron et al. [2016]. The developmental relationship between DHEA and visual attention is mediated by structural plasticity of cortico-amygdalar networks, *Psychoneuroendocrinology, 70:*122–133; with permission from Elsevier.)

awareness. DHEA-related alterations in amygdalar structure and covariance with cortical visual networks may, therefore, play a role in downregulating amygdala-dependent neural bias in visual processing related to affect modulation. A similar relationship was also present for right amygdala-right parietal covariance (i.e., higher DHEA → negative amygdala-parietal covariance → higher visuomotor processing scores on the Wechsler Intelligence Scale for Child (WISC) coding subtest) (Nguyen, Gower et al., 2016). These results are consistent with the role of the parietal cortex in early visuospatial attention and support a role for DHEA in the modulation of amygdala-dependent plasticity of the parietal cortex, perhaps through a downregulation of sensorimotor bias related to affect modulation. Finally, we found significant associations among DHEA, right amygdala-right prefrontal covariance, and attention problems (i.e., higher DHEA → negative amygdala-prefrontal covariance → improved attention) (Nguyen, Gower et al., 2016). Although both occipital and parietal areas have been involved in preconscious, early detection of visual stimuli, the subgenual ACC has been implicated in more global attentional processes and conscious inhibition of emotions. DHEA-related alterations in amygdalar-prefrontal networks may, therefore, contribute to decrease the bias in attentional processing due to affect modulation. In sum, DHEA may uniquely regulate bottom-up amygdala-dependent cortical plasticity and visual attention systems in humans.

Testosterone, Cortico-Limbic Development, and Cognition

Testosterone is thought to play a crucial role in mediating sexual differentiation of brain structures. Previous investigations using experimental animal models documented testosterone-related masculinization of limbic brain areas. In particular, changes in hippocampal volume were shown to be sex specific, with a negative correlation with pubertal maturation in boys and a positive correlation in girls. Several reports from our group and others noted testosterone-dependent, sex-specific changes in gray matter and CTh across puberty in several regions involved in cognitive control and emotional regulation (Nguyen, McCracken, Ducharme, Botteron et al., 2013). Taken together, the current literature suggests that testosterone may alter the structural relationship, or covariance, between the cortex and hippocampus. Studies of testosterone administration or deprivation reported an overall beneficial effect of this hormone on cognitive control and mnemonic processes, at least in adults, which further supports the importance of testosterone in determining brain structure and function.

In this context, our research group investigated the relationship among cortico-hippocampal structural networks, executive function, and verbal memory. We found that prefrontal-hippocampal covariance varied as a function of testosterone levels, but only in boys (see Figure 6.3). Boys also showed a specific association between positive prefrontal-hippocampal covariance (as seen at higher testosterone levels) and lower performance on specific components of executive function; namely, monitoring of the action process and flexibly shifting between actions. Finally, we found the association between testosterone and a specific aspect of executive function (monitoring) to be significantly mediated by prefrontal-hippocampal structural covariance, but we found no effects of testosterone-related structural covariance on verbal memory in either sex. Although studies in younger male samples are scarce, at least one other study did not find any effects of testosterone replacement on executive function in young adult men, particularly with regard to psychomotor speed and the process of switching between cognitive sets (Haren, Wittert, Chapman, Coates, & Morley, 2005). In addition, another study found treatment with testosterone to result in improved verbal memory for men if conversion to estradiol occurred (Cherrier et al., 2005). Finally, our group found levels of aggressive behavior to vary according to testosterone-related prefrontal-amygdala

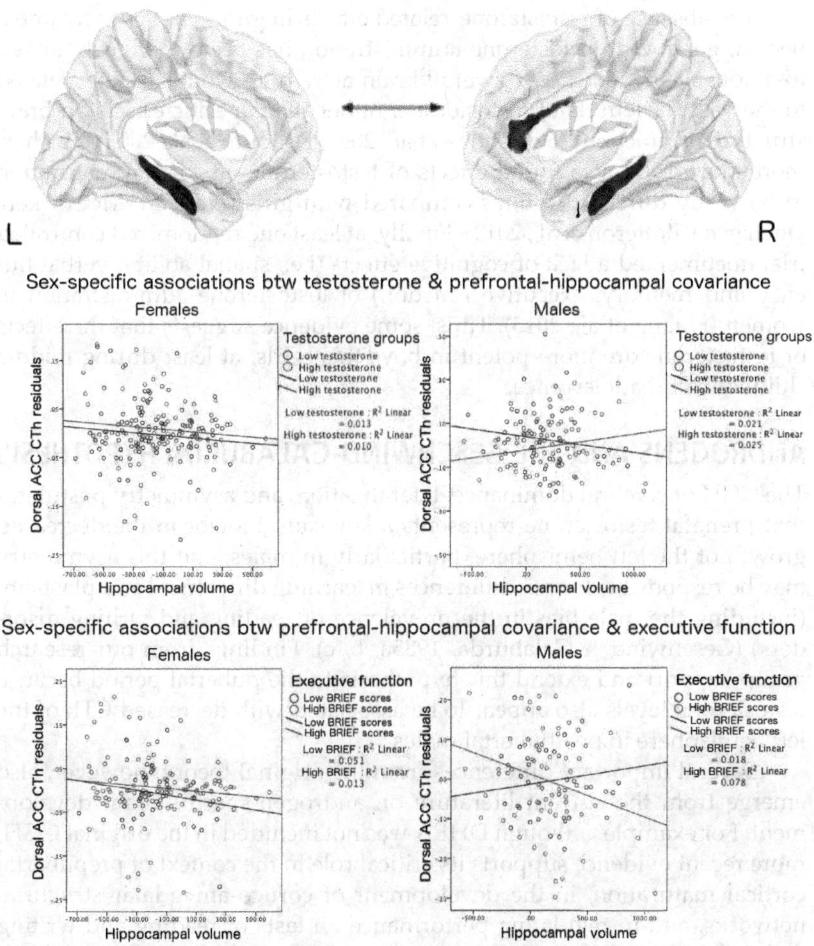

Figure 6.3. This figure shows sex-specific associations among testosterone, prefrontal-hippocampal structural covariance, and executive function. Prefrontal-hippocampal covariance varied in boys according to testosterone levels. In addition, prefrontal-hippocampal structural covariance in boys predicted scores on different components of executive function (Behavior Rating Inventory of Executive Function—Monitor and Shift subscales). (From Nguyen, Lew, Albaugh, Botternon et al. [2017]. Sex-specific associations of testosterone with prefrontal-hippocampal development and executive function, *Psychoneuroendocronology, 76*:206–207; with permission from Elsevier.)

covariance (possibly related to disinhibition and release from top-down inhibition), which may compound the effects of prefrontal-hippocampal covariance on executive function, at least in boys (Nguyen, McCracken et al., 2016). Taken together, these findings suggest that testosterone may have a specific detrimental effect on executive function during middle childhood and adolescence through the alteration of prefrontal-limbic covariance and behavioral activation effects.

The absence of testosterone-related effects in girls is not entirely unexpected, given that testosterone administration has been previously linked to a nonspecific increase in overall brain activity in women (likely related to the vascular effects of testosterone, rather than specific effects on brain structure or function) (Fernandez et al., 2003; Redoute et al., 2005). Furthermore, our group also found effects of testosterone on cortical maturation to be vastly different in boys compared with girls (Nguyen, McCracken, Ducharme, Botteron et al., 2013). Finally, at least one randomized controlled trial documented a lack of cognitive effects (i.e., spatial ability, verbal fluency and memory, executive function) of testosterone administration in women (Huang et al., 2015). Thus, some evidence suggests that the effects of testosterone are more potent in boys than girls, at least during middle childhood and adolescence.

ANDROGENS AND THE GESCHWIND-GALABURDA HYPOTHESIS

The GGH of cerebral dominance, lateralization, and asymmetry postulates that prenatal testosterone represents a key causal factor in the decreased growth of the left hemisphere, particularly in males, and this asymmetry may be responsible for sex differences in learning differences and plasticity (including the male bias in the prevalence of reading and writing disorders) (Geschwind & Galaburda, 1985a, b, c). Findings from our research group support and extend this hypothesis to the pubertal period because testosterone levels also appear to be associated with decreased CTh of the left hemisphere in postpubertal males.

Several important differences from the original theory, however, also emerge from the current literature on androgen-specific brain development. For example, although DHEA was not included in the original GGH, more recent evidence supports its critical role in the context of prepubertal cortical maturation, in the development of cortico-amygdalar structural networks, and in regulating performance on tests of reading and writing abilities. DHEA shows its most rapid rise during middle childhood, coinciding with the rapid development of language. In addition, its production rises during adrenarche, an endocrine event specific to humans and the great apes, suggesting that it may play an important evolutionary role (Campbell, 2006). As such, DHEA may serve to protect certain metabolically active brain regions (including those of the language network) from the neurotoxic effects of glucocorticoids, allowing them to form later, long-range connections with more distal brain areas necessary for the promotion of complex, human-specific skills (Campbell, 2011). These effects of DHEA, however, are independent from sex, suggesting that other unidentified factors may underlie the male vulnerability to developing disorders of reading and writing.

This male vulnerability may be based on a multitude of factors, including, but not limited to, the interactive effects of DHEA and testosterone during middle childhood as well as the interaction between the pubertal effects of DHEA and the preexisting brain phenotype shaped by testosterone and the XY gene complement in utero. It is interesting to note that our research group found that testosterone appears to be more closely associated with global executive/behavior control, particularly in boys, as opposed to more primary abilities such as verbal/language functions; this finding diverges slightly from the original GGH. Still, these effects of testosterone do support a vulnerability model in the context of which boys may show lower executive function related to the effect of this androgen on prefrontal-hippocampal covariance. One could speculate that abnormalities in DHEA levels or DHEA signaling may be more likely to lead to reading/writing disorders during middle childhood in the context of this already vulnerable male brain phenotype.

FUTURE RESEARCH QUESTIONS

Three particular areas of research require clarification with regard to the CNS effects of androgens: 1) the interconversion between steroid hormones and the difficulty in assigning an effect to one specific hormone, 2) differences between the animal CNS and the human CNS, and 3) differences in the effects of steroid hormones, depending on the developmental period under examination.

REFERENCES

Amedi, A., Floel, A., Knecht, S., Zohary, E., & Cohen, L. (2004). Transcranial magnetic stimulation of the occipital pole interferes with verbal processing in blind subjects. *Nature Neuroscience, 7*, 1266–1270.

Amodio, D. M., & Frith, C. D. (2006). Meeting of minds: The medial frontal cortex and social cognition. *Nature Review Neuroscience, 7*(4), 268–277. doi:10.1038/nrn1884

Blakemore, S.-J. (2008). The social brain in adolescence. *Nature Review Neuroscience, 9*(4), 267–277. doi:10.1038/nrn2353

Campbell, B. (2006). Adrenarche and the evolution of human life history. *American Journal of Human Biology, 18*(5), 569–589. doi:10.1002/ajhb.20528

Campbell, B. C. (2011). Adrenarche and middle childhood. *Human Nature: An Interdisciplinary Biosocial Perspective, 22*(3), 327–349. doi:10.1007/s12110-011-9120-x

Cherrier, M. M., Matsumoto, A. M., Amory, J. K., Ahmed, S., Bremner, W., Peskind, E. R.,…Craft, S. (2005). The role of aromatization in testosterone supplementation: Effects on cognition in older men. *Neurology, 64*(2), 290–296.

Crone, E. A., & Dahl, R. E. (2012). Understanding adolescence as a period of social-affective engagement and goal flexibility. *Nature Review Neuroscience, 13*(9), 636–650. doi:10.1038/nrn3313

Fernandez, G., Weis, S., Stoffel-Wagner, B., Tendolkar, I., Reuber, M., Beyenburg, S.,… Elger, C. E. (2003). Menstrual cycle-dependent neural plasticity in the adult human brain is hormone, task, and region specific. *Journal of Neuroscience, 23*(9), 3790–3795.

Fransson, P., & Marrelec, G. (2008). The precuneus/posterior cingulate cortex plays a pivotal role in the default mode network: Evidence from a partial correlation network analysis. *NeuroImage, 42*(3), 1178–1184. doi:10.1016/j.neuroimage.2008.05.059

Geschwind, N., & Galaburda, A. M. (1985a). Cerebral lateralization. Biological mechanisms, associations, and pathology: I. A hypothesis and a program for research. *Archives of Neurology, 42*(5), 428–459.

Geschwind, N., & Galaburda, A. M. (1985b). Cerebral lateralization. Biological mechanisms, associations, and pathology: II. A hypothesis and a program for research. *Archives of Neurology, 42*(6), 521–552.

Geschwind, N., & Galaburda, A. M. (1985c). Cerebral lateralization. Biological mechanisms, associations, and pathology: III. A hypothesis and a program for research. *Archives of Neurology, 42*(7), 634–654.

Grieve, S. M., Korgaonkar, M. S., Clark, C. R., & Williams, L. M. (2011). Regional heterogeneity in limbic maturational changes: Evidence from integrating cortical thickness, volumetric and diffusion tensor imaging measures. *NeuroImage, 55*(3), 868–879. doi:10.1016/j.neuroimage.2010.12.087

Haren, M. T., Wittert, G. A., Chapman, I. M., Coates, P., & Morley, J. E. (2005). Effect of oral testosterone undecanoate on visuospatial cognition, mood and quality of life in elderly men with low-normal gonadal status. *Maturitas, 50*(2), 124–133. doi:10.1016/j.maturitas.2004.05.002

Herting, M. M., Gautam, P., Spielberg, J. M., Dahl, R. E., & Sowell, E. R. (2015). A longitudinal study: Changes in cortical thickness and surface area during pubertal maturation. *PLoS One, 10*(3), e0119774. doi:10.1371/journal.pone.0119774

Huang, G., Wharton, W., Travison, T. G., Ho, M. H., Gleason, C., Asthana, S.,…Basaria, S. (2015). Effects of testosterone administration on cognitive function in hysterectomized women with low testosterone levels: a dose-response randomized trial. *Journal of Endocrinological Investigation, 38*(4), 455–461. doi:10.1007/s40618-014-0213-3

Maninger, N., Wolkowitz, O. M., Reus, V. I., Epel, E. S., & Mellon, S. H. (2009). Neurobiological and neuropsychiatric effects of dehydroepiandrosterone (DHEA) and DHEA sulfate (DHEAS). *Frontiers of Neuroendocrinology, 30*(1), 65–91. doi:10.1016/j.yfrne.2008.11.002

Morris, J. S., DeGelder, B., Weiskrantz, L., & Dolan, R. J. (2001). Differential extrageniculostriate and amygdala responses to presentation of emotional faces in a cortically blind field. *Brain, 124*, 1241–1252. doi:10.1093/Brain/124.6.1241

Nelson, C. A., Monk, C. S., Lin, J., Carver, L. J., Thomas, K. M., & Truwit, C. L. (2000). Functional neuroanatomy of spatial working memory in children. *Developmental Psychology, 36*(1), 109–116. doi:10.1037/0012-1649.36.1.109

Nguyen, T. V., Gower, P., Albaugh, M. D., Botteron, K. N., Hudziak, J. J., Fonov, V. S., . . . McCracken, J. T. (2016). The developmental relationship between DHEA and visual attention is mediated by structural plasticity of cortico-amygdalar networks. *Psychoneuroendocrinology, 70*, 122–133. doi:10.1016/j.psyneuen.2016.05.003

Nguyen, T. V., Lew, J., Albaugh, M. D., Botteron, K. N., Hudziak, J. J., Fonov, V. S., . . . McCracken, J. T. (2017). Sex-specific associations of testosterone with prefrontal-hippocampal development and executive function. *Psychoneuroendocrinology, 76*, 206–217.

Nguyen, T. V., McCracken, J. T., Albaugh, M. D., Botteron, K. N., Hudziak, J. J., & Ducharme, S. (2016). A testosterone-related structural brain phenotype predicts aggressive behavior from childhood to adulthood. *Psychoneuroendocrinology, 63*, 109–118. doi:10.1016/j.psyneuen.2015.09.021

Nguyen, T. V., McCracken, J., Ducharme, S., Botteron, K. N., Mahabir, M., Johnson, W., . . . Karama, S. (2013). Testosterone-related cortical maturation across childhood and adolescence. *Cerebral Cortex, 23*(6), 1424–1432. doi:10.1093/cercor/bhs125

Nguyen, T. V., McCracken, J. T., Ducharme, S., Cropp, B. F., Botteron, K. N., Evans, A. C., & Karama, S. (2013). Interactive effects of dehydroepiandrosterone and testosterone on cortical thickness during early brain development. *Journal of Neuroscience, 33*(26), 10840–10848. doi:10.1523/JNEUROSCI.5747-12.2013

Pasley, B. N., Mayes, L. C., & Schultz, R. T. (2004). Subcortical discrimination of unperceived objects during binocular rivalry. *Neuron, 42*(1), 163–172. doi:10.1016/S0896-6273(04)00155-2

Pegna, A. J., Khateb, A., Lazeyras, F., & Seghier, M. L. (2005). Discriminating emotional faces without primary visual cortices involves the right amygdala. *Nature Neuroscience, 8*(1), 24–25. doi:10.1038/nn1364

Redoute, J., Stoleru, S., Pugeat, M., Costes, N., Lavenne, F., Le Bars, D.,…Pujol, J. F. (2005). Brain processing of visual sexual stimuli in treated and untreated hypogonadal patients. *Psychoneuroendocrinology, 30*(5), 461–482. doi:10.1016/j.psyneuen.2004.12.003

Remer, T., Boye, K. R., Hartmann, M. F., & Wudy, S. A. (2005). Urinary markers of adrenarche: Reference values in healthy subjects, aged 3–18 years. *Journal of Clinical Endocrinology and Metabolism, 90*(4), 2015–2021. doi:10.1210/jc.2004-1571

Saxe, R. R., Whitfield-Gabrieli, S., Scholz, J., & Pelphrey, K. A. (2009). Brain regions for perceiving and reasoning about other people in school-aged children. *Child Development, 80*(4), 1197–1209. doi:10.1111/j.1467-8624.2009.01325.x

Szaflarski, J. P., Schmithorst, V. J., Altaye, M., Byars, A. W., Ret, J., Plante, E., & Holland, S. K. (2006). A longitudinal functional magnetic resonance imaging study of language development in children 5 to 11 years old. *Annals of Neurology, 59*(5), 796–807. doi:10.1002/ana.20817

Vasconsuelo, A., Pronsato, L., Ronda, A. C., Boland, R., & Milanesi, L. (2011). Role of 17β-estradiol and testosterone in apoptosis. *Steroids, 76*(12), 1223–1231. doi:10.1016/j.steroids.2011.08.001

Vuilleumier, P., Armony, J. L., Driver, J., & Dolan, R. J. (2003). Distinct spatial frequency sensitivities for processing faces and emotional expressions. *Nature Neuroscience, 6*(6), 624–631. doi:10.1038/nn1057

Williams, M. A., Morris, A. P., McGlone, F., Abbott, D. F., & Mattingley, J. B. (2004). Amygdala responses to fearful and happy facial expressions under conditions of binocular suppression. *Journal of Neuroscience, 24*(12), 2898–2904. doi:10.1523/JNEUROSCI.4977-03.2004

INTEGRATIVE SUMMARY 1

Brain Development, Hormones, and Immunology and the Geschwind-Galaburda Hypothesis

Margaret M. McCarthy

SUMMARY

Discerning the origins of neurological, neuropsychiatric, and neurodegenerative disorders is a central goal of brain science. The importance of early life developmental events to all disorders of the nervous system is increasingly evident. This importance was presaged in the Geschwind-Galaburda Hypothesis (GGH), which sought to connect the disparate observations of combined higher rates of dyslexia, left-handedness, and enhanced immune activation, all of which appeared more frequently in males than females. The observation that the left cortical hemisphere was smaller than the right in these individuals to a greater degree than normally seen was the first clue that events during development were the source of the connection between these endpoints. But what precisely happens during development has remained vexingly difficult to unravel. New research points to the potential impact of stress experienced by the mother during pregnancy on her developing fetus, as well as the enduring impact of inflammation. Puberty is a critical yet unappreciated dynamic period of brain development, and events at this time can enhance and dampen earlier developmental perturbations. Finally, the brain develops in a body, and the body affects the brain. Although the brain was once thought to be a relatively static organ, it is a constantly changing, adapting, and responsive system that is more malleable than ever imagined. Discovering what biological processes permit or prevent brain plasticity will provide the tools to enhance recovery and even promote regeneration following acute injury or chronic damage. The GGH provides a firm framework on which those processes can be elucidated.

COMMENTARY

Corrupting the words of the famous geneticist Dobszhansky, "Nothing in biology makes sense except in the light of DEVELOPMENT," captures the essence of the GGH (Geschwind & Galaburda, 1985a, b, c), which attempts to unravel the puzzle of higher rates of dyslexia, left-handedness, and enhanced immune activation in males compared with females. In this volume, Galaburda clarifies that the hypothesis is really a theory and generates multiple testable hypotheses. The essential theory is that elevated testosterone activity in developing male fetuses delays development of the left cerebral hemisphere, resulting in cortical asymmetry that is greater in males than females and, in turn, underlies sex differences in left-handedness, dyslexia, and other language disorders. The theory also loosely postulates that this imbalance connects in some way with altered immune function in people who have dyslexia and their family members. Each of the authors in Section II: Brain Development, Hormones, and Immunology reexamine the hypothesis through the lens of current understanding and highlight how impactful and yet still poorly understood the theory has been and continues to be.

Development is a simple word for a dynamic multifactorial robust-yet-sensitive-to-perturbation compilation of multiple processes. What happens first affects what happens next, which affects what happens next, and so on in an increasingly constrained one-way trajectory. Identifying the parameters limiting yet driving normal development is essential to identifying nodes of vulnerability to perturbation and what factors gain access to those nodes. The developmental trajectory associated with the GGH is a prototypical example of both sides of the coin as it attempts to identify the normal parameters and how they are perturbed.

The origins of asymmetry in bilateral organisms have long been a mystery, but the existence of these asymmetries is well documented across a wide range of features from the directionality of internal organs to wingspan in birds. Rubenstein (see Chapter 3) reviews recent breakthrough findings indicating that as early as gastrulation and the formation of the neural plate in vertebrate embryos, the leftward beating of cilia from cells in the midline node are the origins of left/right body asymmetry. This provides a clear proof of principle that asymmetry can have genetic origins, and Rubenstein goes on to assert that genetics is a key entry point into understanding the biological origins of dyslexia. Intriguingly, two genes identified to date are central to synthesis of estradiol, an essential hormone in brain development, and cilia function, reminding us that the origins of a complex language disorder can be both unexpected and distant from the identified malformation.

In line with this view, De Vries reminds us that the brain develops in a body, and the body can affect the brain (see Chapter 4). Sex differences in the brain are predominantly determined developmentally and the Organizational-Activational Hypothesis of sexual differentiation of the brain parallels the GGH, and its premise is that elevated testosterone from the fetal male testis organizes the developing brain along a masculine phenotype that is then activated by testosterone in adulthood following puberty (Phoenix, Goy, Gerall, & Young, 1959). This hypothesis is similar to the GGH in that it is really a theory based on observations and has fostered many and varied hypotheses. These hypotheses have increased in sophistication in recent years, including the paradigm-shifting inclusion of the sex chromosomes as potential sources of neural sex differences (Arnold et al., 2004, 2016; McCarthy & Arnold, 2011). De Vries was one of several investigators to exploit the power of the 4-core-genotype mouse model in which the impact of the gonads can be isolated from sex chromosome complement (De Vries et al., 2002). This is achieved by transferring the gene on the Y chromosome that directs testis development to an autosome that allows for XX individuals to develop a testis, whereas XY individuals without the gene will develop ovaries. De Vries et al. used this model and found that sexually differentiated endpoints that are directly relevant to reproduction are largely driven by gonadal steroids, whereas more cognitive or emotional endpoints have some influence from the sex chromosomes, although this does not mean to say there is no hormonal effect. This complex combination of genetics and hormones is most apt for the GGH because the sex differences in cortical asymmetry cannot be entirely explained by either sex chromosome complement or testosterone (see Chapter 1).

Early life programming, also called the Barker Hypothesis (Barker, 1995), is a third overarching hypothesis, yet really a theory, highlighted in this section. This hypothesis is based on observation and postulates that what happens during development imprints on the developing organism, including the brain, and the individual carries that imprint across the life span. In Chapter 5, O'Connor and colleagues review the existing epidemiological and preclinical animal literature on maternal anxiety and its impact on developing fetuses. Evidence for the loss of cortical asymmetry and the induction of it is presented and varies by brain region, species, nature of the stressor, and study. The authors rightly conclude that the current collection of findings neither supports nor refutes a role for prenatal stress in GGH, but that is in large part due to not framing studies in the context of GGH. Doing so is a worthy goal going forward, particularly considering a potential role for inflammation, which is becoming increasingly linked to an unhealthy prenatal environment (Najjar, Pearlman, Alper, Najjar, & Devinsky, 2013). The reasons why there is a hyper-responsive immune profile in left handers and those with dyslexia remains mysterious, and the prenatal stress model could provide much needed insight.

One of the challenges of development is that it goes on for a long time, especially in humans. Indeed, the extended juvenile/adolescent period in humans is unique among primates. Nguyen focuses on puberty, which is itself an extended developmental process that begins with adrenarche and the production of adrenal steroids (see Chapter 6). The GGH emphasized in utero testosterone, but Nguyen points out the dehydroepiandrosterone and its sulfated by-product are also androgens with a myriad of complex effects. Moreover, many of those effects may be counter to those of testosterone, and they may differ in males and females to modulate cortical and amygdala asymmetries. Puberty is emerging as a new sensitive period in brain development but remains poorly understood and underinvestigated (Herbison, 2007; Sisk & Foster, 2004). Higher cognitive brain regions are still immature as puberty is entered and continue to develop once puberty is complete, thus progressing through a developmental juggernaut. Language and reading abilities are well established by this time, but the potential that other functionally important brain asymmetries are either established, exaggerated, or eliminated as a result of this passage is realistic. This expands the impact and significance of the GGH to a new stage of brain development and provides additional frameworks on which new hypotheses can be formulated and tested.

In conclusion, the GGH remains as exciting and vital a theory as when it was first proposed in the *Archives of Neurology* in the mid-1980s. It has stood the test of time and testing, with some aspects being supported and others debated, refined, and retested. Integrating it with other large theoretical frameworks such as the Organizational-Activational Hypothesis of sexual differentiation of the brain and the Barker Hypothesis (Barker, 1995) has the potential to advance these constructs by highlighting commonalities and novelties. Rigorous hypothesis testing in preclinical and animal models can best proceed when grounded in a robust theoretical framework supported by both observational and empirical data; the GGH provides that framework.

REFERENCES

Arnold, A. P., Reue, K., Eghbali, M., Vilain, E., Chen, X., Ghahramani, N.,…Williams-Burris, S. M. (2016). The importance of having two X chromosomes. *Philosophical Transactions of the Royal Society of London. Series B, Biological Sciences, 371,* 20150113.

Arnold, A. P., Xu, J., Grisham, W., Chen, X., Kim, Y. H., & Itoh, Y. (2004). Minireview: Sex chromosomes and brain sexual differentiation. *Endocrinology, 145,* 1057–1062.

Barker, D. J. (1995). Fetal origins of coronary heart disease. *British Medical Journal, 311*(6998), 171–174.

De Vries, G. J., Rissman, E. F., Simerly, R. B., Yang, L. Y., Scordalakes, E. M., Auger, C. J.,… Arnold, A. P. (2002). A model system for study of sex chromosome effects on sexually dimorphic neural and behavioral traits. *Journal of Neuroscience, 22*(20), 9005–9014.

Dobszhansky, T. (1973). Nothing in biology makes sense except in the light of evolution. *The American Biology Teacher, 35*(3), 125–129.

Geschwind, N., & Galaburda, A. M. (1985a). Cerebral lateralization. Biological mechanisms, associations, and pathology: I. A hypothesis and a program for research. *Archives of Neurology,* 42(5), 428–459.

Geschwind, N., & Galaburda, A. M. (1985b). Cerebral lateralization. Biological mechanisms, associations, and pathology: II. A hypothesis and a program for research. *Archives of Neurology,* 42(6), 521–552.

Geschwind, N., & Galaburda, A. M. (1985c). Cerebral lateralization. Biological mechanisms, associations, and pathology: III. A hypothesis and a program for research. *Archives of Neurology,* 42(7), 634–654.

Herbison, A. E. (2007). Genetics of puberty. *Hormone Research, 68*(Suppl 5), 75–79.

McCarthy, M. M., & Arnold, A. P. (2011). Reframing sexual differentiation of the brain. *Nature Neuroscience, 14,* 677–683.

Najjar, S., Pearlman, D. M., Alper, K., Najjar, A., & Devinsky, O. (2013). Neuroinflammation and psychiatric illness. *Journal of Neuroinflammation, 10*(43), 816. doi:10.1186/1742-2094-10-43

Phoenix, C. H., Goy, R. W., Gerall, A. A., & Young, W. C. (1959). Organizing action of prenatally administered testosterone proprionate on the tissues mediating mating behavior in the female guinea pig. *Endocrinology, 65,* 369–382.

Sisk, C. L., & Foster, D. L. (2004). The neural basis of puberty and adolescence. *Nature Neuroscience, 7,* 1040–1047.

SECTION III

Sex Differences

CHAPTER 7

Brain Asymmetries and Sex Differences in Developmental Dyslexia

Franck Ramus, Irene Altarelli,
Katarzyna Jednoróg, Jingjing Zhao, and Lou Scotto di Covella

SUMMARY

Thirty years ago, Geschwind and Galaburda (1985a, b, c) hypothesized complex links among genetic factors, prenatal environment, sex, brain asymmetries, and the susceptibility to developmental disorders, in particular, developmental dyslexia. This chapter reviews studies our team conducted on the neuroanatomy of developmental dyslexia. Among other results, we replicated Galaburda's original finding of an altered asymmetry of the planum temporale (PT) in dyslexia by using magnetic resonance imaging (MRI) in a large group of children. This difference in asymmetry, however, was only found in boys. We observed that the cortical region dedicated to visual word recognition (located in the left hemisphere) is thicker in readers who do not have dyslexia than in readers who have dyslexia, but only among girls. We also found that two white matter tracts connecting posterior and anterior regions of the brain and participating in reading show different hemispheric asymmetry patterns in children with dyslexia and control children. Finally, we discovered a difference in the asymmetry of the depth of the central sulcus (a major fold of the brain) between children with dyslexia and control children, with a different pattern in boys and in girls. Overall, we found that when individuals with dyslexia and control individuals differ in brain anatomy, the differences depend on the cerebral hemisphere, and they are not the same in males and females. In other words, brain asymmetry seems key to understanding the neuroanatomical basis of dyslexia, and this neuroanatomical basis seems to partly differ between the sexes. Several possible explanations are discussed that are consistent with Geschwind and Galaburda's original ideas.

INTRODUCTION

Geschwind and Galaburda (1985a, b, c) hypothesized complex links among genetic factors, prenatal (particularly hormonal) environment, sex, brain asymmetries, and the susceptibility to developmental disorders, in particular developmental dyslexia. Thirty years later, it is of great interest to go back to the original hypotheses and ask which have stood the test of time.

This chapter looks at sex, brain asymmetries, and dyslexia. (Geschwind and Galaburda also hypothesized additional links, such as the immune system and handedness, which are not addressed here.)

Developmental dyslexia is a specific impairment in the acquisition of reading skills that is not solely accounted for by mental age, visual acuity problems, or inadequate schooling (World Health Organization, 2011). Cases with dyslexia have been described for more than a century, and the hypothesis of a specific brain dysfunction was present from the first case (Morgan, 1896). Investigations into the neuroanatomical bases of developmental dyslexia began with a postmortem case description by Drake (1968), followed by another series of postmortem cases by Galaburda and colleagues (Galaburda & Eidelberg, 1982; Galaburda & Kemper, 1979; Galaburda & Livingstone, 1993; Galaburda, Menard, & Rosen, 1994; Galaburda, Sherman, Rosen, Aboitiz, & Geschwind, 1985; Humphreys, Kaufmann, & Galaburda, 1990). Neuroimaging was soon used to broaden the investigations to live individuals and, therefore, to significantly increase the number of cases studied. The first brain scans of individuals with dyslexia may be attributed to Hier, LeMay, Rosenberger, and Perlo (1978) and exploded from the 1990s with the advent of MRI, starting with Rumsey et al. (1986).

Postmortem dissections had the exceptional advantage of offering both macro- and microscopic views of the dyslexic brain and used standard histological techniques for microscopic views, thus giving some insight into cellular and architectonic phenomena. Disruptions observed in such studies included ectopic neurons, heterotopias, and dysplasias, converging toward the hypothesis of a disturbance of neuronal migration (Drake, 1968; Galaburda et al., 1985; Humphreys et al., 1990). Furthermore, disruptions were predominantly localized in left perisylvian regions, suggesting a straightforward link with language and reading networks (Ramus, 2004). Although this hypothesis is extremely seductive, it has suffered from its reliance on a small number of cases (a total of 4 males and 3 females with dyslexia, compared with 13 male and 3 female controls) and from the lack of additional direct confirmation since those initial observations. Although MRI has allowed researchers to reassess macroscopic observations on a much larger scale, it has until now lacked the contrast and resolution that would be necessary to observe minute disruptions of neuronal migration. Therefore, modern neuroimaging has left that initial hypothesis largely untouched, but it has multiplied the opportunities of observing macroscopic alterations of brain structure by using different types of image acquisition and a growing variety of image analysis techniques. The following reviews investigations from our group that turned out to be relevant with respect to Geschwind and Galaburda's hypotheses and to Galaburda and colleagues' postmortem discoveries.

ASYMMETRY OF THE PLANUM TEMPORALE

The PT is an associative auditory area located on the upper surface of the superior temporal gyrus, posterior to Heschl's gyrus. The surface area of this region is larger in the left than in the right hemisphere in about 65% of the general population (Geschwind & Levitsky, 1968). The idea that the asymmetry of the PT was altered in developmental dyslexia emerged as early as the 1980s, following postmortem observations on seven dyslexic brains (Galaburda et al., 1985; Humphreys et al., 1990). Many subsequent (in vivo) MRI studies, however, failed to provide clear evidence for or against PT anomalies in dyslexia (Best & Demb, 1999; Bloom, Garcia-Barrera, Miller, Miller, & Hynd, 2013; Heiervang et al., 2000; Larsen, Høien, Lundberg, & Ødegaard, 1990; Robichon, Bouchard, Demonet, & Habib, 2000; Semrud-Clikeman et al., 1996). Among the 20 MRI studies on the topic published thus far, only four reported symmetry or rightward asymmetry of the PT in developmental dyslexia (Bloom et al., 2013; Hynd, Semrud-Clikeman, Lorys, Novey, & Eliopulos, 1990; Larsen et al., 1990; Semrud-Clikeman et al., 1996).

Albert Galaburda discussed these results in person and mentioned that most of these studies had measured the PT area differently from the original postmortem studies (personal communication, May 2, 2008). Differences primarily concerned the criteria defining the anterior boundary of the PT region and its posterior boundary, both of which can have an impact on measured asymmetry and influence observed group differences (Altarelli et al., 2014). The reconstruction techniques applied to the MRI images and the measurements used is a second methodological factor of importance. For example, some studies compared the length of the PT between participants with dyslexia and control participants measured from a limited number of sagittal slices, whereas others analyzed its surface area, sometimes taking convolutions into account. Finally, a third relevant factor regards the limited number of participants typically studied (15 per group on average), with heterogeneous criteria for inclusion in the dyslexic group (for instance, as related to the presence or absence of comorbidities, the evaluation of nonverbal IQ, the types of reading tests used, and cut-off criteria applied). In addition, groups have not always been matched for relevant variables such as age, nonverbal IQ, and handedness.

Taking the previous information into account, we asked Galaburda to teach us to manually delineate and measure the PT using the same criteria he used in postmortem studies to control our work. We did so in 81 children (a population more than twice as large as those typically studied in previous reports), with groups carefully matched for age, sex, nonverbal IQ, and handedness, comparing different anatomical criteria. We obtained a partial confirmation of Galaburda and colleagues' (1985) initial findings in the sense that we found that boys with dyslexia tended to have rightward asymmetrical PT surface, when control boys showed the expected leftward asymmetry (Altarelli et al., 2014).

Furthermore, we found that the set of anatomical criteria used is crucial and, as suggested earlier, is likely to explain at least some of the previous discordant findings. Moreover, we uncovered that sex is another key factor to consider because it was found to interact with the diagnosis of dyslexia, such that PT surface area asymmetry was disrupted only in boys with dyslexia but not in girls with dyslexia. These results are in line with the observation of different types of cortical anomalies in the postmortem brains of men and women with dyslexia (Galaburda et al., 1985; Humphreys et al., 1990).

CORTICAL THICKNESS OF THE VISUAL WORD FORM AREA

We also conducted a study of cortical thickness (CTh) and surface area in two independent datasets comprising age-matched and sex-matched children with dyslexia and control children (87 participants in total). Whole-brain analyses did not yield reliable differences. We focused, however, on functionally defined regions of interest in the ventral occipito-temporal cortex. We used functional MRI and localized the individual peaks of response to houses, faces, and words and investigated CTh differences between children with dyslexia and control children around those peaks. We found a specific reduction in CTh around the peak of response to words in the left hemisphere in girls with dyslexia compared with control girls. This result was found to be consistent in a second independent age-matched dataset as well as in a comparison of children with dyslexia with reading-matched controls. Although the interpretation of differences in CTh remains unclear, this difference did not seem to be entirely due to differences in reading achievement. No such result was observed in boys, however (Altarelli et al., 2013). This was our first encounter with a basic sex difference in the neuroanatomical basis of dyslexia.

WHITE MATTER AND DIFFUSION IMAGING

We applied diffusion imaging and spherical deconvolution tractography to reconstruct four white matter tracts relevant to reading and language—the inferior frontal-occipital fasciculus (IFOF), the inferior longitudinal fasciculus (ILF) (Catani & Thiebaut de Schotten, 2008), the three segments of the superior longitudinal fasciculus (SLF I: dorsal, SLF II: middle, SLF III: ventral; Thiebaut de Schotten et al., 2011), and the three segments of the arcuate fasciculus (AF: long fronto-temporal, posterior temporo-parietal, and anterior fronto-parietal segments; Catani, Jones, & ffytche, 2005). The main group difference that we found was that children with dyslexia showed increased hindrance modulated orientational anisotropy (HMOA) in the right SLF compared with age-matched controls. Our main results, however, are best demonstrated as asymmetry patterns in which children with dyslexia and control children differ in the lateralization of two tracts—IFOF and SLF II. Children with dyslexia show a weaker left-lateralization (of HMOA) of IFOF than control children and a right-lateralization of SLF II (in which control children show no hemispheric difference in SLF II), and these deviant lateralization patterns are related to variations in reading performance in children with dyslexia (Zhao, Thiebaut de Schotten, Altarelli, Dubois, & Ramus, 2016).

GYRIFICATION PATTERNS

Beyond total brain volume, brain gyrification is the most visible source of anatomical differences between individuals. Furthermore, it occurs during early brain development and is relatively stable after birth (Cachia et al., 2016; Mangin et al., 2004; White, Su, Schmidt, Kao, & Sapiro, 2010), making it a prime target for the study of early markers of cognitive function and impairment. Finally, sulci are reliable (although certainly imperfect) landmarks commonly used to identify cytoarchitectonic brain regions in the absence of direct histological evidence. The shape, localization, and size of sulci have been rarely investigated in the study of individual differences and neurodevelopmental disorders, despite their great potential relevance (Borst et al., 2016; Cachia et al., 2008, 2014; Gay et al., 2013; Plaze et al., 2009).

As far as we are aware, only two studies investigated potential differences in gyrification patterns in dyslexia. Im and collaborators (Im, Raschle, Smith, Grant, & Gaab, 2015) used a complex graph-theoretical analysis to investigate global sulcal patterns across the temporo-parietal and occipito-temporal regions. They reported atypical sulcal patterns in the dyslexic and the at-risk prereader groups, characterized by more sulcal basins of smaller size than in the control groups.

We adopted an entirely different approach and focused on quantitative and qualitative properties of three semimanually labelled sulci of interest—the central sulcus, the sylvian fissure, and the superior temporal sulcus—in a population of 142 children with dyslexia and 106 control children across three countries (Scotto di Covella et al., submitted). One result stood out concerning the mean depth of the central sulcus, in the form of a triple interaction among hemisphere, sex, and group. We found the central sulcus is deeper in the right hemisphere in control girls but symmetrical in control boys. It was deeper in the left hemisphere in girls with dyslexia and deeper in the right in boys with dyslexia, however. No other difference was found.

These studies are quite unique and deserve to be replicated before conclusions are drawn. Furthermore, the functional interpretation of the observed gyrification differences remains entirely open. Nevertheless, we think that the study of cortical sulci is a promising research avenue that deserves more attention than it has received.

DISCUSSION

Neuroanatomical differences have typically been observed in only one hemisphere throughout the dyslexia literature—usually the left one, which is associated with language and reading functions in most individuals, but sometimes the right one, which is often interpreted as a form of right-hemisphere compensation for left-hemisphere deficiencies (Richardson & Price, 2009). Hardly any bilateral difference has been reported. We find that when the two hemispheres are systematically compared, however, not only do group differences emerge in one or the other hemisphere, but some differences also emerge as differences in sheer lateralization patterns (e.g., our studies on the PT, on diffusion imaging, and on sulci). This suggests that, beyond the fact that reading is predominantly localized in the left hemisphere, developmental dyslexia may be more fundamentally related to a disruption of normal brain lateralization as hypothesized by Geschwind and Galaburda (1985a).

Although the uneven sex ratio in dyslexia (about 2 boys for 1 girl) has been known for a long time, sex has never been a factor of much interest in dyslexia research, whether at the cognitive or neural level. Early neuroimaging studies either focused exclusively on males (e.g., Brown et al., 2001; Paulesu et al., 2001) or included too few male and female participants to afford splitting groups by sex. Even larger studies often neglected to do so or still had too few participants of one sex (e.g., Pernet, Poline, Demonet, & Rousselet, 2009; Tamboer, Scholte, & Vorst, 2015). Yet, the few studies that had enough participants of each sex and deliberately entered sex as a factor in their analyses have often found that the brain differences between participants with dyslexia and control participants differ depending on sex. This has been reported in a voxel-based morphometry study (Evans, Flowers, Napoliello, & Eden, 2014), our study on the CTh of the visual word form area (Altarelli et al., 2013), our study of the PT (Altarelli et al., 2014), and our study of cortical sulci (Scotto di Covella et al., submitted). It is also worth recalling that Galaburda and colleagues' initial postmortem observations also differed to some extent between male and female brains (Galaburda et al., 1985; Humphreys et al., 1990).

Thus, indications show that the neural basis of developmental dyslexia may differ between males and females to some extent. The reasons are not fully known, although a role for fetal sex hormones may seem plausible (Geschwind & Galaburda, 1985a). According to one scenario, female hormones would play a protective role with respect to various disruptions of brain development. Thus, females may require more severe brain disruptions to show an equal level of cognitive dysfunction as males (Ramus, 2006), a prediction consistent with animal studies (Fitch, Brown, Tallal, & Rosen, 1997; Hall, Pazara, & Linseman, 1991; Roof, Duvdevani, Braswell, & Stein, 1994), postmortem work on dyslexia (Galaburda et al., 1985; Humphreys et al., 1990), and the study showing a group difference in the CTh of the visual word form area in females only (Altarelli et al., 2013). Some neuroimaging studies, however, have also reported differences exclusively in males (Altarelli et al., 2014) or different patterns in males and females (Evans et al., 2014; Scotto di Covella et al., submitted), thus suggesting that the picture is more complex than a simple severity gradient.

The study of sex differences in dyslexia is still in its infancy. It deserves more attention than it has attracted so far, however, and it should be investigated more systematically in all dyslexia studies that have enough participants of each sex. Indeed, disruptions of brain development operate against a genetic and hormonal background that differs between the sexes, which sometimes leads to basic sex differences in the neuroanatomical properties of interest. For instance, we have serendipitously discovered sex differences in the asymmetry of Heschl's gyrus surface (Altarelli et al., 2014) and of central sulcus depth in our studies on the neuroanatomy of dyslexia (Scotto di Covella et al., submitted). More generally, the importance of not neglecting sex anymore is not restricted to dyslexia research, but it has recently been recognized for all of neuroscience (Cahill, 2006) and medical research (Clayton & Collins, 2014).

REFERENCES

Altarelli, I., Leroy, F., Monzalvo, K., Fluss, J., Billard, C., Dehaene-Lambertz, G.,…Ramus, F. (2014). Planum temporale asymmetry in developmental dyslexia: Revisiting an old question. *Human Brain Mapping, 35,* 5717–5735.

Altarelli, I., Monzalvo, K., Iannuzzi, S., Fluss, J., Billard, C., Ramus, F., & Dehaene-Lambertz, G. (2013). A functionally guided approach to the morphometry of occipitotemporal regions in developmental dyslexia: Evidence for differential effects in boys and girls. *Journal of Neuroscience, 33,* 11296–11301.

Best, M., & Demb, J. B. (1999). Normal planum temporale asymmetry in dyslexics with a magnocellular pathway deficit. *Neuroreport, 10*(3), 607–612.

Bloom, J. S., Garcia-Barrera, M. A., Miller, C. J., Miller, S. R., & Hynd, G. W. (2013). Planum temporale morphology in children with developmental dyslexia. *Neuropsychologia, 51,* 1684–1692.

Borst, G., Cachia, A., Tissier, C., Ahr, E., Simon, G., & Houdé, O. (2016). *Early cerebral constraints on reading skills in school-age children: An MRI study.* Retrieved from http://doi.org/10.1111/mbe.12098

Brown, W. E., Eliez, S., Menon, V., Rumsey, J. M., White, C. D., & Reiss, A. L. (2001). Preliminary evidence of widespread morphological variations of the brain in dyslexia. *Neurology, 56,* 781–783.

Cachia, A., Borst, G., Tissier, C., Fisher, C., Plaze, M., Gay, O.,…Raznahan, A. (2016). Longitudinal stability of the folding pattern of the anterior cingulate cortex during development. *Developmental Cognitive Neuroscience, 19,* 122–127.

Cachia, A., Borst, G., Vidal, J., Fischer, C., Pineau, A., Mangin, J.-F., & Houdé, O. (2014). The shape of the ACC contributes to cognitive control efficiency in preschoolers. *Journal of Cognitive Neuroscience, 26,* 96–106.

Cachia, A., Paillère-Martinot, M.-L., Galinowski, A., Januel, D., de Beaurepaire, R., Bellivier, F.,…Duchesnay, E. (2008). Cortical folding abnormalities in schizophrenia patients with resistant auditory hallucinations. *NeuroImage, 39,* 927–935.

Cahill, L. (2006). Why sex matters for neuroscience. *Nature Reviews Neuroscience, 7,* 477–484.

Catani, M., Jones, D. K., & ffytche, D. H. (2005). Perisylvian language networks of the human brain. *Annals of Neurology, 57,* 8–16.

Catani, M., & Thiebaut de Schotten, M. (2008). A diffusion tensor imaging tractography atlas for virtual in vivo dissections. *Cortex, 44,* 1105–1132. doi:10.1016/j.cortex.2008.05.004

Clayton, J. A., & Collins, F. S. (2014). NIH to balance sex in cell and animal studies. *Nature, 509*(7500), 282–283.

Drake, W. E. (1968). Clinical and pathological findings in a child with a developmental learning disability. *Journal of Learning Disabilities, 1,* 486–502.

Evans, T. M., Flowers, D. L., Napoliello, E. M., & Eden, G. F. (2014). Sex-specific gray matter volume differences in females with developmental dyslexia. *Brain Structure and Function, 219*, 1041–1054.

Fitch, R. H., Brown, C. P., Tallal, P., & Rosen, G. D. (1997). Effects of sex and MK-801 on auditory-processing deficits associated with developmental microgyric lesions in rats. *Behavior Neuroscience, 111*, 404–412.

Galaburda, A. M., & Eidelberg, D. (1982). Symmetry and asymmetry in the human posterior thalamus: II. Thalamic lesions in a case of developmental dyslexia. *Archives of Neurology, 39*, 333–336.

Galaburda, A. M., & Kemper, T. L. (1979). Cytoarchitectonic abnormalities in developmental dyslexia: A case study. *Annals of Neurology, 6*, 94–100.

Galaburda, A. M., & Livingstone, M. (1993). Evidence for a magnocellular defect in developmental dyslexia. *Annals of the New York Academy of Sciences, 682*, 70–82.

Galaburda, A. M., Menard, M. T., & Rosen, G. D. (1994). Evidence for aberrant auditory anatomy in developmental dyslexia. *Proceedings of the National Academy of Science USA, 91*, 8010–8013.

Galaburda, A. M., Sherman, G. F., Rosen, G. D., Aboitiz, F., & Geschwind, N. (1985). Developmental dyslexia: Four consecutive patients with cortical anomalies. *Annals of Neurology, 18*, 222–233.

Gay, O., Plaze, M., Oppenheim, C., Mouchet-Mages, S., Gaillard, R., Olié, J.-P.,…Cachia, A. (2013). Cortex morphology in first-episode psychosis patients with neurological soft signs. *Schizophrenia Bulletin, 39*, 820–829.

Geschwind, N., & Galaburda, A. M. (1985a). Cerebral lateralization. Biological mechanisms, associations, and pathology: I. A hypothesis and a program for research. *Archives of Neurology, 42*(5), 428–459.

Geschwind, N., & Galaburda, A. M. (1985b). Cerebral lateralization. Biological mechanisms, associations, and pathology: II. A hypothesis and a program for research. *Archives of Neurology, 42*(6), 521–552.

Geschwind, N., & Galaburda, A. M. (1985c). Cerebral lateralization. Biological mechanisms, associations, and pathology: III. A hypothesis and a program for research. *Archives of Neurology, 42*(7), 634–654.

Geschwind, N., & Levitsky, W. (1968). Human brain: Left-right asymmetries in temporal speech region. *Science, 161*, 186–187.

Hall, E. D., Pazara, K. E., & Linseman, K. L. (1991). Sex differences in postischemic neuronal necrosis in gerbils. *Journal of Cerebral Blood Flow Metabolism, 11*, 292–298.

Heiervang, E., Hugdahl, K., Steinmetz, H., Inge Smievoll, A., Stevenson, J., Lund, A.,… Lundervold, A. (2000). Planum temporale, planum parietale and dichotic listening in dyslexia. *Neuropsychologia, 38*, 1704–1713.

Hier, D. B., LeMay, M., Rosenberger, P. B., & Perlo, V. P. (1978). Developmental dyslexia: Evidence for a subgroup with a reversal of cerebral asymmetry. *Archives of Neurology, 35*, 90–92.

Humphreys, P., Kaufmann, W. E., & Galaburda, A. M. (1990). Developmental dyslexia in women: Neuropathological findings in three patients. *Annals of Neurology, 28*, 727–738.

Hynd, G. W., Semrud-Clikeman, M., Lorys, A. R., Novey, E. S., & Eliopulos, D. (1990). Brain morphology in developmental dyslexia and attention deficit disorder/hyperactivity. *Archives of Neurology, 47*, 919–926.

Im, K., Raschle, N. M., Smith, S. A., Grant, P., & Gaab, N. (2015). *Atypical sulcal pattern in children with developmental dyslexia and at-risk kindergarteners.* Retrieved from http://doi.org/10.1093/cercor/bhu305

Larsen, J. P., Høien, T., Lundberg, I., & Ødegaard, H. (1990). MRI evaluation of the size and symmetry of the planum temporale in adolescents with developmental dyslexia. *Brain and Language, 39*, 289–301.

Mangin, J. F., Riviere, D., Cachia, A., Duchesnay, E., Cointepas, Y., Papadopoulos-Orfanos, D.,…Regis, J. (2004). A framework to study the cortical folding patterns. *NeuroImage, 23*(Suppl 1), S129-138.

Morgan, W. P. (1896). A case of congenital word blindness. *British Medical Journal, 2*, 1378.

Paulesu, E., Démonet, J. F., Fazio, F., McCrory, E., Chanoine, V., Brunswick, N.,…Frith, U. (2001). Dyslexia: Cultural diversity and biological unity. *Science, 291*, 2165–2167.

Pernet, C. R., Poline, J. B., Demonet, J. F., & Rousselet, G. A. (2009). Brain classification reveals the right cerebellum as the best biomarker of dyslexia. *BMC Neuroscience, 10,* 67.

Plaze, M., Paillère-Martinot, M.-L., Penttilä, J., Januel, D., de Beaurepaire, R., Bellivier, F.,... Cachia, A. (2009). "Where do auditory hallucinations come from?" A brain morphometry study of schizophrenia patients with inner or outer space hallucinations. *Schizophrenia Bulletin, 37*(1), 212–221.

Ramus, F. (2004). Neurobiology of dyslexia: A reinterpretation of the data. *Trends in Neurosciences, 27,* 720–726.

Ramus, F. (2006). A neurological model of dyslexia and other domain-specific developmental disorders with an associated sensorimotor syndrome. In G. D. Rosen (Ed.), *The dyslexic brain: New pathways in neuroscience discovery* (pp. 75–101). Mahwah, NJ: Lawrence Erlbaum Associates.

Richardson, F. M., & Price, C. J. (2009). Structural MRI studies of language function in the undamaged brain. *Brain Structure and Function, 213,* 511–523. doi:10.1007/s00429-009-0211-y

Robichon, F., Bouchard, P., Demonet, J., & Habib, M. (2000). Developmental dyslexia: Re-evaluation of the corpus callosum in male adults. *European Neurology, 43,* 233–237.

Roof, R. L., Duvdevani, R., Braswell, L., & Stein, D. G. (1994). Progesterone facilitates cognitive recovery and reduces secondary neuronal loss caused by cortical contusion injury in male rats. *Experimental Neurology, 129,* 64–69.

Rumsey, J. M., Dorwart, R., Vermess, M., Denckla, M. B., Kruesi, M. J., & Rapoport, J. L. (1986). Magnetic resonance imaging of brain anatomy in severe developmental dyslexia. *Archives of Neurology, 43,* 1045–1046.

Scotto di Covella, L., Leroy, F., Cauvet, E., Girard, F., Couvignou, M., Caron-Guyon, J.,... Ramus, F. (submitted). Morphometry of three sulci in developmental dyslexia.

Semrud-Clikeman, M., Hooper, S. R., Hynd, G. W., Hern, K., Presley, R., & Watson, T. (1996). Prediction of group membership in developmental dyslexia, attention deficit hyperactivity disorder, and normal controls using brain morphometric analysis of magnetic resonance imaging. *Archives of Clinical Neuropsychology, 11*(6), 521–528.

Tamboer, P., Scholte, H. S., & Vorst, H. C. (2015). Dyslexia and voxel-based morphometry: Correlations between five behavioural measures of dyslexia and gray and white matter volumes. *Annals of Dyslexia, 65*(3), 121–141.

Thiebaut de Schotten, M., Dell'Acqua, F., Forkel, S. J., Simmons, A., Vergani, F., Murphy, D. G. M., & Catani, M. (2011). A lateralized brain network for visuospatial attention. *Nature Neuroscience, 14,* 1245–1246.

White, T., Su, S., Schmidt, M., Kao, C.-Y., & Sapiro, G. (2010). The development of gyrification in childhood and adolescence. *Brain and Cognition, 72*(1), 36–45. http://doi.org/10.1016/j.bandc.2009.10.009

World Health Organization. (2011). *International statistical classification of diseases and related health problems: 10th revision* (4th ed.). Geneva, Switzerland: Author.

Zhao, J., Thiebaut de Schotten, M., Altarelli, I., Dubois, J., & Ramus, F. (2016). Altered hemispheric lateralization of white matter tracts in developmental dyslexia: Evidence from spherical deconvolution tractography. *Cortex, 76,* 51–62.

CHAPTER 8

The Sexual Dimorphism of the Human Brain

Discriminating Between Effects of Brain Size and Effects of Sex Independent of Brain Size

Eileen Luders

SUMMARY

Neuroimaging studies have revealed a number of striking sex differences in the anatomy of the human brain. Yet, because male brains are usually bigger than female brains overall, the question arises as to whether any of the observed sex differences are simply a by-product of brain size or whether they constitute sex effects entirely independent of brain size. The goal of this chapter is to convey that the latter assumption is true for some but certainly not all findings. The scientific basis for this conclusion was drawn using a unique set of 96 brain scans (48 male, 48 female) in which 24 male and 24 female brains were matched pair-wise for their size. When comparing these two matched subsamples, the local amount of gray matter was found to be larger in female brains than male brains in specific brain regions. In contrast, there were no sex differences with respect to the thickness of the corpus callosum, which is the brain's largest white matter structure. Matching for gross brain size is a valuable approach to clarify whether observed sex differences in regional brain anatomy exist independently of brain size effects or whether they are simply an artifact thereof.

INTRODUCTION

Sex-specific brain anatomy is a classic and contemporary subject of interest for laymen and researchers alike. The larger weight and volume of male brains compared with female brains is the most replicated and robust finding across scientific studies (Giedd, Raznahan, Mills, & Lenroot, 2012; Gong, He, & Evans, 2011; Luders & Toga, 2010; Sacher, Neumann, Okon-Singer, Gotowiec, & Villringer, 2013). Interestingly, this substantial sex difference is only partly accounted for by the larger body dimensions in men (Ankney, 1992; Peters, 1991). In addition to sex differences in overall (global) brain size, traditional postmortem as well as modern in vivo studies have revealed several other, more region-specific (local) differences between male and female brains. Such local sex differences seem to exist when measuring the surface area, thickness or gyrification of the cerebral cortex; the shapes, volumes, and areas of subcortical structures; or the local amount of gray and white matter tissue, as detailed in countless original research studies as well as systematic reviews and meta-analyses (Bishop & Wahlsten, 1997; Cosgrove, Mazure, & Staley, 2007; Eliot, 2011; Giedd et al., 2012; Gong et al., 2011; Luders & Toga, 2010; Tan, Ma, Vira, Marwha, & Eliot, 2016).

THE UNANSWERED QUESTION(S)

Little consensus exists among scientists as to which brain regions are truly sexually dimorphic, even with more than 40 years of fervently developing and refining neuroimaging techniques and image analysis tools. In fact, the presence, magnitude, and direction of observed sex differences seems to vary depending not only on the brain structure examined (e.g., cerebral cortex *vs.* corpus callosum), but also on the particular brain feature analyzed (e.g., cortical thickness *vs.* cortical gyrification) and the degree of regional specificity (e.g., total callosal area *vs.* segment-specific callosal area).

However, the largest source of discrepancy seems to be whether regional measurements are adjusted (corrected) for individual brain size or not. That is, studies comparing male and female brains without applying brain size corrections often reveal larger regional measures in males than females (or no sex differences). In contrast, studies comparing male and female brains applying brain size corrections often reveal larger regional measures in females than males (or no sex differences). Nevertheless, even in the latter case, findings seem to vary greatly depending on the nature of the brain size correction (e.g., covarying, residualization, data scaling, ratio estimation). Unfortunately, there is still no agreement among scientists about the most suitable correction approach (Barnes et al., 2010; Bermudez & Zatorre, 2001; Smith, 2005), and the mere process of correcting may introduce artificial effects. For example, overcorrections in smaller (female) brains could result in falsely inflated measures, especially if allometric relationships exist between brain size and the size of its substructures, which could lead to the impression of larger regional volumes in smaller (female) brains.

So, ultimately, the question arises whether any sex differences in the macroanatomy of the human brain, as observed in scientific studies, are simply a by-product of sex-typical brain size or of false corrections thereof. Studying men and women who are very similar or even equal in brain size might provide an answer to this question because brain size (or any correction thereof) would no longer play a significant role and, thus, not affect study outcomes. Given that male and female brains usually differ in size by about 10%–15% on average (Cosgrove et al., 2007; Giedd et al., 2012; Gur et al., 1999; Leonard et al., 2008; Luders et al., 2005), some caution is warranted with respect to generalizing the findings because such brain-size matched samples may be considered sex atypical in terms of brain size (i.e., male brains might be smaller than what is common; female brains might be larger than what is common). Nevertheless, studying brain size-matched samples is a valuable approach to unscramble the differential impact of size and sex because any observed male–female difference would constitute a sex effect entirely independent of brain size. However, compiling brain size-matched samples is rather challenging, given the unique (uncommon) sex-specific brain sizes required. Fortunately, scientists have started building enormous digital databases containing reams of structural (T1-weighted) magnetic resonance imaging brain scans. Such large-scale databases constitute an opportunity to select from a vast pool of individuals (and thus brain sizes) and tackle the aforementioned question, especially as information on sex is usually available together with the digital brain data.

GETTING SOME ANSWERS: FOUR UNIQUE SUBSAMPLES—TWO SELECTED STUDIES

This chapter summarizes the outcomes of a research project using 96 structural brain scans carefully selected from the International Consortium for Brain Mapping (ICBM) database (Mazziotta et al., 2009). Specifically, this set of data (n = 96) contained two main samples—a matched sample (n = 48) with 24 male and 24 female brains identical in size and an extreme sample (n = 48) with 24 extremely large (XL) male and 24 extremely small (XS) female brains (see Table 8.1). Characteristics of these four subsamples, as well as details on the image acquisition, participant selection, and calculation of the individual brain volumes, are provided in the original publications (Luders, Gaser, Narr, & Toga, 2009; Luders, Toga, & Thompson, 2014). We compared male and female brains from these four compiled subsamples—altogether using four separate contrasts (see Figure 8.1)—to address the question of whether anatomical differences between male and female brains exist independently of brain size effects or whether brain size accounts for observed sex differences. We systematically decreased the mean brain volume difference between males and females (i.e., gray panel *vs.* black panel), starting from 402 ml (Contrast 1), to 217 ml (Contrast 2), to 185 ml (Contrast 3), to 0 ml (Contrast 4) (see Figure 8.1). We used this stepwise approach and conducted two separate studies. Study I focused on the gray matter, specifically voxel-wise gray matter volume (Luders et al., 2009), and Study II focused on the corpus callosum (a white matter structure), specifically point-wise callosal thickness (Luders et al., 2014).

Table 8.1. Subsample-specific brain volumes

	Matched sample (n = 48)		Extreme sample (n = 48)	
	Matched Males (n = 24)	Matched Females (n = 24)	Extreme (XL) Males (n = 24)	Extreme (XS) Females (n = 24)
Brain volume	1,406.57 ± 101.69	1,406.62 ± 101.41	1,623.74 ± 68.94	1,221.58 ± 62.99

Brain volume (mean ± standard deviation) is shown in milliliters.

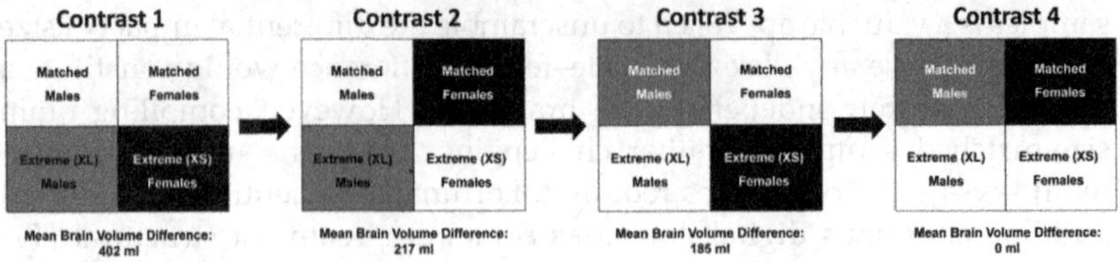

Figure 8.1. The four subgroups defined by brain size and sex and the four different comparisons (Contrasts 1–4). Fields in gray (males) and in black (females) indicate the two groups contrasted. The mean brain volume difference between males and females was systematically decreased.

Study I: Voxel-wise Gray Matter

The applied analysis stream in Study I is commonly known as voxel-based morphometry and has been detailed elsewhere (Ashburner & Friston, 2000). Briefly, the structural brain scans from the database undergo several automated processing steps to generate bias-corrected, spatially normalized, modulated, and smoothed individual gray matter segments. Please refer to the original publication for Study I for further details pertaining to data processing as well as the subsequent statistical analyses (Luders et al., 2009).

The voxel-wise gray matter volume was always smaller in males than in females (males < females) when comparing the four subgroups defined by brain size and sex (see Figure 8.2). Interestingly, the larger the sex difference in brain size, the more pronounced the sex difference in voxel-wise gray matter (see Contrasts 1–4). On one hand, the gray matter sex difference may reflect a size effect in that smaller brains in general (independent of sex) contain more gray matter locally so that there are the same numbers of processing units as in larger brains. On the other hand, the gray matter sex difference may be a result of (over) correcting brains during the normalization to a normative template with fixed dimensions in which small (female) brains get larger and large (male) brains get smaller. To explore these options, although not further described in this chapter, Study I also tested for main effects of brain size (matched males *vs.* XL males; matched females *vs.* XS females) and sex-by-size interactions. However, none of these analyses revealed any significant effects. Thus, brain size or (false) corrections thereof cannot be the driving force for the observed gray matter sex

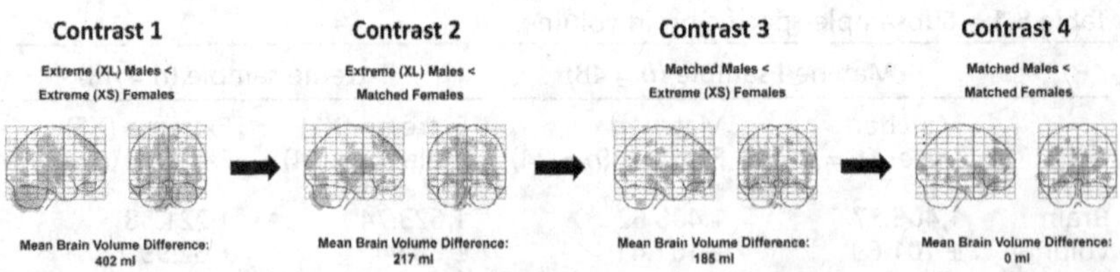

Figure 8.2. Significant gray matter differences between male and female brains. When comparing the four subgroups defined by brain size and sex, the voxel-wise gray matter volume was always smaller in males than in females (males < females).

difference. Perhaps even more relevant is that there is still more gray matter in female brains when comparing matched males and matched females (see Contrast 4), although the effect is slightly less pronounced than in the other contrasts. Thus, we may conclude that the sex differences observed in Contrast 4 constitute sex effects independent of brain size.

Study II: Point-wise Callosal Thickness

The applied analysis stream in Study II is commonly known as the callosal thickness approach and has been detailed elsewhere (Luders, Narr, Zaidel, Jancke et al., 2006). Briefly, structural brain scans are automatically processed to generate bias-corrected and spatially normalized images, which are then used to outline the corpus callosum in each brain's midsagittal section. Ultimately, the local callosal thickness is calculated at 100 equidistant points. Please refer to the original publication accompanying Study II for more information pertaining to these computations as well as the subsequent statistical analyses (Luders et al., 2009).

The corpus callosum was either thicker in males than in females (males > females) or there were no sex differences at all when comparing the four subgroups defined by brain size and sex (see Figure 8.3). In general, the larger the sex difference in brain size, the more pronounced the sex difference in point-wise callosal thickness (see Contrasts 1–4). More specifically, the male corpus callosum was thicker than the female corpus callosum in almost all sections when comparing the XL males and XS females (see Contrast 1). Reducing the brain size difference between males and females also reduced the observable sex difference in callosal thickness, however, with a few scattered effects remaining in Contrast 2, but none in Contrast 3 or Contrast 4, when comparing the two matched subsamples. Thus, we may conclude that the observed sex differences in callosal thickness (if any) do not constitute size-independent sex effects but rather are driven by brain size.

Studies I and II: Integrative Discussion

The outcomes of the two studies provide important insights into whether sex differences in regional brain anatomy exist independently of brain size effects or whether they are simply a by-product thereof. More important, the study outcomes signify that one rule does not fit all, and the question

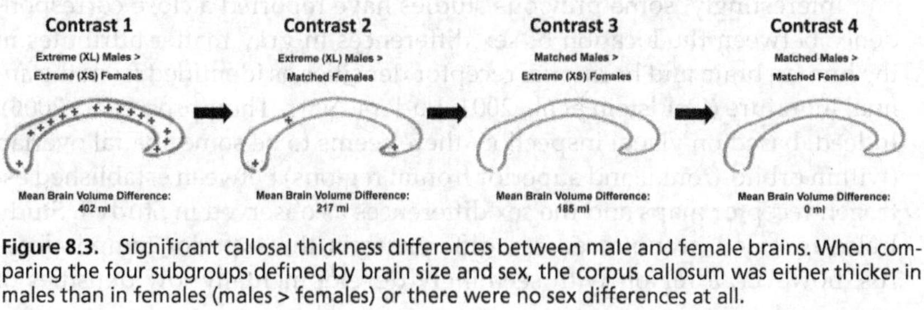

Figure 8.3. Significant callosal thickness differences between male and female brains. When comparing the four subgroups defined by brain size and sex, the corpus callosum was either thicker in males than in females (males > females) or there were no sex differences at all.

of whether brain size or sex per se is the driving force needs to be resolved for each brain attribute separately. More specifically, Study I suggests that the observed male < female differences in voxel-wise gray matter constitute mostly sex effects independent of brain size (but note that the sexual dimorphism decreases slightly when samples are matched for brain size). In contrast, Study II supports the assumption that the observed male > female differences in point-wise callosal thickness are brain size effects; there are no sexually dimorphic regions in the corpus callosum whatsoever when samples are matched for brain size. Thus, it is imperative to avoid any oversimplification and generalization when examining the sexual dimorphism of the brain and attempting to discriminate between brain size effects and sex effects independent of brain size.

RELATIONSHIP TO THE GESCHWIND-GALABURDA HYPOTHESIS

Sex hormones are one of the determinants for sex differences in the brain (Geschwind & Galaburda, 1985a, b, c), and they may not only have growth-promoting and neuroprotective effects but also growth-inhibiting and deleterious effects. More specifically, sex hormones may stimulate angiogenesis, neurogenesis, or synaptogenesis as well as synaptic pruning or cell apoptosis (Arnold & Gorski, 1984; Breedlove, 1994; Kawata, 1995). This, in turn, may affect the local amount of brain tissue and contribute to voxel-wise gray matter as analyzed in Study I. The sex differences (females > males) observed in this study appear relatable to the Geschwind-Galaburda Hypothesis (GGH), especially when focusing on the matched sample in which possible effects of brain size can be excluded with certainty. In terms of the specific locations, which are described in more detail in the original article (Luders et al., 2009), the most pronounced sex differences were located in three main clusters spanning across different brain areas. The largest cluster included the left and right caudate (together with other areas of the basal ganglia) as well as the left orbito-frontal region. The second cluster comprised areas of the left superior temporal gyrus, and the third cluster encompassed areas of the left superior frontal gyrus. The prominence of female > male differences within the left hemisphere is particularly interesting in the framework of the GGH because Geschwind and Galaburda (1985a, b, c) suggested delayed growth of the left hemisphere in male brains due to elevated levels of fetal testosterone.

Interestingly, some previous studies have reported a close correspondence between the location of sex differences in gray matter attributes in the human brain and hormonal receptor densities as identified from the animal literature (Goldstein et al., 2001; Luders, Narr, Thompson et al., 2006). Indeed, based on visual inspection, there seems to be some spatial overlap (within orbito-frontal and superior frontal regions) between established estrogen receptor maps and the sex differences as observed in Study I. Study I also revealed pronounced sex differences in the superior temporal gyrus, however, a region with seemingly developmentally low densities of estrogen receptors (Goldstein et al., 2001). Thus, further research is needed because sex steroid receptor maps identified in animals do not necessarily translate to humans, lack data on specific brain regions, or only convey information on receptor density for a specific sex steroid. Moreover, aside from activating or organizing effects of sex hormones, various other mechanisms and determinants are likely to cause, sustain, accentuate, or attenuate the sexual dimorphism of the human brain, as discussed in detail elsewhere (De Vries & Forger, 2015; Eliot, 2011; Luders & Toga, 2010). Future studies carefully characterizing the relationship between observable gray matter via neuroimaging and its actual underlying microstructure, as well as determinants for the sexual dimorphism of the human brain (e.g., the modulating effects of hormones as predicted by the GGH), are clearly necessary.

CONCLUSION

Ample evidence exists for a sexual dimorphism in the anatomy of the human brain. Because male brains are usually bigger than female brains, the question arises whether observed sex differences are simply a sequel to size differences. The goal of this chapter was to convey that this assumption is true for some but not all findings. For example, the local amount of gray matter was found to be larger in female brains regardless of brain size, albeit such size-independent sex effects were evident only in specific regions (rather than evenly distributed across the brain). A few other studies have conducted similar systematic investigations to determine the interplay between sex and brain size and/or the impact of data scaling on analysis outcomes (Ardekani, Figarsky, & Sidtis, 2013; Giedd et al., 2012; Jancke, Staiger, Schlaug, Huang, & Steinmetz, 1997; Leonard et al., 2008; Luders, Narr, Thompson et al., 2006; Luders, Narr, Zaidel, Thompson, & Toga, 2006; Sowell et al., 2007). Brain size matching as described in this chapter is a suitable approach to clarify whether observed sex differences are attributable to brain size or to biological sex per se. Further research—varying not only the brain structure of interest but also its analyzed features and degree of regional specificity—is certainly warranted to gain further insights and paint a more complete and differentiated picture.

REFERENCES

Ankney, C. D. (1992). The brain size/IQ debate. *Nature, 360*(6402), 292.

Ardekani, B. A., Figarsky, K., & Sidtis, J. J. (2013). Sexual dimorphism in the human corpus callosum: An MRI study using the OASIS brain database. *Cerebral Cortex, 23*(10), 2514–2520. doi:10.1093/cercor/bhs253

Arnold, A. P., & Gorski, R. A. (1984). Gonadal steroid induction of structural sex differences in the central nervous system. *Annual Review. of Neuroscience, 7*, 413–442.

Ashburner, J., & Friston, K. J. (2000). Voxel-based morphometry: The methods. *Neuroimage, 11*(6), 805–821.

Barnes, J., Ridgway, G. R., Bartlett, J., Henley, S. M., Lehmann, M., Hobbs, N.,...Fox, N. C. (2010). Head size, age and gender adjustment in MRI studies: A necessary nuisance? *NeuroImage, 53*(4), 1244–1255. doi:10.1016/j.neuroimage.2010.06.025

Bermudez, P., & Zatorre, R. J. (2001). Sexual dimorphism in the corpus callosum: Methodological considerations in MRI morphometry. *NeuroImage, 13*(6 Pt 1), 1121–1130. doi:10.1006/nimg.2001.0772

Bishop, K. M., & Wahlsten, D. (1997). Sex differences in the human corpus callosum: Myth or reality? *Neuroscience and Biobehavioral Review, 21*(5), 581–601.

Breedlove, S. M. (1994). Sexual differentiation of the human nervous system. *Annual Review of Psychology, 45,* 389–418.

Cosgrove, K. P., Mazure, C. M., & Staley, J. K. (2007). Evolving knowledge of sex differences in brain structure, function, and chemistry. *Biological Psychiatry, 62*(8), 847–855.

De Vries, G. J., & Forger, N. G. (2015). Sex differences in the brain: A whole body perspective [Review]. *Biology of Sex Differences, 6,* 15. doi:10.1186/s13293-015-0032-z

Eliot, L. (2011). The trouble with sex differences [Comparative Study Review]. *Neuron, 72*(6), 895–898. doi:10.1016/j.neuron.2011.12.001

Geschwind, N., & Galaburda, A. M. (1985a). Cerebral lateralization. Biological mechanisms, associations, and pathology: I. A hypothesis and a program for research. *Archives of Neurology, 42*(5), 428–459.

Geschwind, N., & Galaburda, A. M. (1985b). Cerebral lateralization. Biological mechanisms, associations, and pathology: II. A hypothesis and a program for research. *Archives of Neurology, 42*(6), 521–552.

Geschwind, N., & Galaburda, A. M. (1985c). Cerebral lateralization. Biological mechanisms, associations, and pathology: III. A hypothesis and a program for research. *Archives of Neurology, 42*(7), 634–654.

Giedd, J. N., Raznahan, A., Mills, K. L., & Lenroot, R. K. (2012). Review: Magnetic resonance imaging of male/female differences in human adolescent brain anatomy. *Biology of Sex Differences, 3*(1), 19. doi:2042-6410-3-19 [pii] 10.1186/2042-6410-3-19

Goldstein, J. M., Seidman, L. J., Horton, N. J., Makris, N., Kennedy, D. N., Caviness, V. S., Jr.,...Tsuang, M. T. (2001). Normal sexual dimorphism of the adult human brain assessed by in vivo magnetic resonance imaging. *Cerebral Cortex, 11*(6), 490–497.

Gong, G., He, Y., & Evans, A. C. (2011). Brain connectivity: Gender makes a difference. *Neuroscientist, 17*(5), 575–591. doi:1073858410386492 [pii] 10.1177/1073858410386492

Gur, R. C., Turetsky, B. I., Matsui, M., Yan, M., Bilker, W., Hughett, P., & Gur, R. E. (1999). Sex differences in brain gray and white matter in healthy young adults: Correlations with cognitive performance. *Journal of Neuroscience, 19*(10), 4065–4072.

Jancke, L., Staiger, J. F., Schlaug, G., Huang, Y., & Steinmetz, H. (1997). The relationship between corpus callosum size and forebrain volume. *Cerebral Cortex, 7*(1), 48–56.

Kawata, M. (1995). Roles of steroid hormones and their receptors in structural organization in the nervous system. *Neuroscience Research, 24*(1), 1–46.

Leonard, C. M., Towler, S., Welcome, S., Halderman, L. K., Otto, R., Eckert, M. A., & Chiarello, C. (2008). Size matters: Cerebral volume influences sex differences in neuroanatomy. *Cerebral Cortex, 18*(12), 2920–2931.

Luders, E., Gaser, C., Narr, K. L., & Toga, A. W. (2009). Why sex matters: Brain size independent differences in gray matter distributions between men and women. *Journal of Neuroscience, 29*(45), 14265–14270.

Luders, E., Narr, K. L., Thompson, P. M., Rex, D. E., Woods, R. P., DeLuca, H.,...Toga, A. W. (2006). Gender effects on cortical thickness and the influence of scaling. *Human Brain Mapping, 27*(4), 314–324.

Luders, E., Narr, K. L., Thompson, P. M., Woods, R. P., Rex, D. E., Jancke, L.,...Toga, A. W. (2005). Mapping cortical gray matter in the young adult brain: Effects of gender. *NeuroImage, 26*(2), 493–501.

Luders, E., Narr, K. L., Zaidel, E., Thompson, P. M., Jancke, L., & Toga, A. W. (2006). Parasagittal asymmetries of the corpus callosum. *Cerebral Cortex, 16*(3), 346–354.

Luders, E., Narr, K. L., Zaidel, E., Thompson, P. M., & Toga, A. W. (2006). Gender effects on callosal thickness in scaled and unscaled space. *Neuroreport, 17*(11), 1103–1106.

Luders, E., & Toga, A. W. (2010). Sex differences in brain anatomy. *Progress Brain Research, 186,* 3–12. doi:B978-0-444-53630-3.00001-4 [pii] 10.1016/B978-0-444-53630-3.00001-4

Luders, E., Toga, A. W., & Thompson, P. M. (2014). Why size matters: Differences in brain volume account for apparent sex differences in callosal anatomy. *NeuroImage, 84,* 820–824. doi:10.1016/j.neuroimage.2013.09.040

Mazziotta, J. C., Woods, R., Iacoboni, M., Sicotte, N., Yaden, K., Tran, M.,…Toga, A. W. (2009). The myth of the normal, average human brain: The ICBM experience: (1) subject screening and eligibility. *Neuroimage, 44*(3), 914–922.

Peters, M. (1991). Sex differences in human brain size and the general meaning of differences in brain size. *Canadian Journal of Psychology, 45*(4), 507–522.

Sacher, J., Neumann, J., Okon-Singer, H., Gotowiec, S., & Villringer, A. (2013). Sexual dimorphism in the human brain: Evidence from neuroimaging. *Magnetic Resonance Imaging, 31*(3), 366–375. doi:S0730-725X(12)00217-2 [pii] 10.1016/j.mri.2012.06.007

Smith, R. J. (2005). Relative size versus controlling for size: Interpretation of ratios in research on sexual dimorphism in the human corpus callosum. *Current Anthropology, 46*(2), 249–273.

Sowell, E. R., Peterson, B. S., Kan, E., Woods, R. P., Yoshii, J., Bansal, R.,…Toga, A. W. (2007). Sex differences in cortical thickness mapped in 176 healthy individuals between 7 and 87 years of age. *Cerebral Cortex, 17*(7), 1550–1560.

Tan, A., Ma, W., Vira, A., Marwha, D., & Eliot, L. (2016). The human hippocampus is not sexually-dimorphic: Meta-analysis of structural MRI volumes. *NeuroImage, 124*(Pt A), 350–366. doi:10.1016/j.neuroimage.2015.08.050

CHAPTER 9

Sex Differences in Cognition and Learning

David S. Hong

SUMMARY

The role of sex in learning disorders (LDs) and underlying neurocognitive domains has been vigorously debated. The most recent evidence, however, suggests that there are sex differences in these areas. Although the topic is controversial, improving our understanding of sex-specific processes may provide critical insights into neurobiological mechanisms underlying developmental dyslexia and how best to treat it. LDs arise early in development and are more often found in males, whereas pathologies that arise during the pubertal period tend to be more predominant in females. Both genetics and hormones have been explored to explain this phenomenon, and a large body of work has defined how sex hormones (androgens in particular) exert masculinizing effects on brain regions, triggering sex differences. Less research has been conducted on genetic mechanisms, despite evidence that they directly influence neurodevelopment. It remains challenging to discriminate effects of hormonal and genetic influence in humans, given that these processes often interact and exert their influence during similar periods of development. One fruitful model is sex chromosome aneuploidies (SCAs), in which individuals (e.g., those with Turner syndrome [TS] or Klinefelter syndrome [KS]) have an abnormal number of sex chromosomes. Both include unique cognitive profiles: Girls with TS have 50% higher prevalence of math LDs, and boys with KS have 50%–75% higher rates of language-based disorders. There appear to be dose-dependent relationships with the number of sex chromosomes (and associated gene expression) associated with impairments in visuospatial function, language-based difficulties, and structural neuroanatomical differences. Although there is not a parsimonious model integrating sex chromosome function and developmental dyslexia, mounting evidence shows that sex plays a central role in neurodevelopment, cognitive phenotype, and aberrant trajectories of learning (e.g., LDs). Further studies are needed to examine these complex relationships, improve our understanding of LDs, and provide tangible targets of intervention for individuals affected with these conditions.

INTRODUCTION

Studies examining distribution of LDs between males and females have been mixed, with some demonstrating equivalent rates between sexes (Moll, Kunze, Neuhoff, Bruder, & Schulte-Korne, 2014; Shaywitz, Shaywitz,

Fletcher, & Escobar, 1990), whereas others indicate sex-specific differences in prevalence. Indeed, studies indicate a greater than 2:1 sex ratio, with males being disproportionately affected (Hawke, Olson, Willcut, Wadsworth, & DeFries, 2009; Rutter et al., 2004) and male-predominance being further amplified with increasing severity of reading difficulties (Hawke, Wadsworth, Olson, & DeFries, 2007; Olson, 2002). It is tempting to argue that these differences result solely from selection biases arising from sociocultural influences; however, a fairly broad literature base examining sex differences in typical learners provides additional convergent evidence that sex represents a distinct biological factor that independently affects cognitive development. The contentious nature of this topic has led studies in this domain to be hotly debated, with strong arguments for and against sex differences in academic achievement at the population level both in math and sciences or language and reading ability (Hedges & Nowell, 1995; Hyde, 2016). Similarly, research on sex differences in underlying neurocognitive domains have been reported and challenged, with the most commonly discussed trends highlighting increased social and verbal functions in females (Hedges & Nowell, 1995; Willingham & Cole, 1997) and visuospatial ability in males (Voyer, Voyer, & Bryden, 1995).

Although data on math and science achievement may be more equivocal, the most robust effect sizes have been reported on sex differences in verbal ability. This also holds true when examining comparative data across societies—examination of reading literacy in various nations demonstrated substantial effect sizes universally in favor of girls for language-based subjects both at the fourth-grade level and also at 15 years of age (Ogle et al., 2003; Stoet & Geary, 2013), thus highlighting the importance of sex differences research, particularly in the domain of reading. It is understandable why the debate on sex differences in learning results in such heated discourse, given the potential impact and subsequent bias this issue may have on perceived ability in the classroom and later societal outcomes. Neglecting potential sex-specific aspects of learning development is equally problematic, however, if, in fact, sex mediates neurobiological processes underlying cognitive functions. Therefore, improving our understanding of sex-specific processes may provide critical insights into neurobiological mechanisms underlying developmental dyslexia while also raising potential future avenues of treatment for youth affected by LDs.

THE ROLE OF SEX IN NEURODEVELOPMENT AND THE GESCHWIND-GALABURDA HYPOTHESIS

It is well established that LDs represent disorders of neurodevelopment, given their emergence in early childhood. Similarly, LDs are often comorbid with other common neurodevelopmental disorders, such as attention-deficit/hyperactivity disorder, which demonstrates a more pronounced sex-specific skew, ranging up to an order of 3:1 in male-to-female

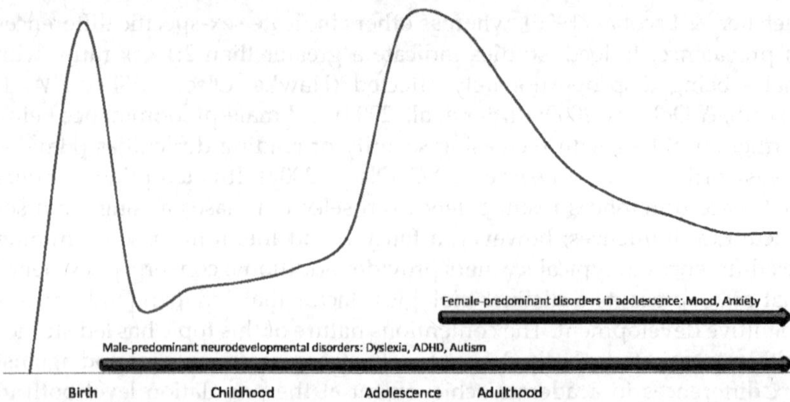

Figure 9.1. Schematic of sex hormone levels over the life span. Neurodevelopmental disorders that present early in development demonstrate male predominance, whereas a number of neuropsychiatric disorders emerging in the pubertal period exhibit female predominance.

prevalence (Wolraich et al., 2014). The consistent male predominance for neurodevelopmental disorders emerging in childhood contrasts starkly with pathologies emerging during the pubertal period, which tend to exhibit female-predominant prevalence (see Figure 9.1).

It is important to understand what neurobiological mechanisms underlie these sex differences. In fact, a number of hypotheses have been presented, spanning molecular, genetic, and hormonal mechanisms, but often converging on the masculinizing effects of testosterone in utero and during the perinatal period. The Geschwind-Galaburda Hypothesis (GGH; 1985a, b, c) outlined a unified framework for this model predicated on a highly programmed set of asymmetric cerebral patterning emerging from this period. This theorem is based on a set of observations, including rates of left-handedness, functional loss in lesion studies, and postmortem findings of cerebral lateralization, that suggest asymmetry is increased in neurodevelopmental disorders, such as developmental dyslexia. Regarding an underlying mechanism, the GGH stated:

> The simplest theory of lateralization is that patterns of asymmetry are strongly determined genetically. Our thesis is that while genes contribute importantly, many influences that lie outside of the gene pool of altered lateralization patterns. The most powerful factors are variations in the chemical environment in fetal life and to a lesser extent in infancy and early childhood. (Geschwind & Galaburda, 1985a, p. 431)

Indeed, much of sex differences research has similarly focused on these two dimensions of genetics and hormones, with a greater proportion of work investigating the effects of sex hormones. A guiding hypothesis incorporating the time-sensitive aspects of this work is defined through the organizational-activational axis, which posits that brain development follows a highly consistent and programmed pattern of changes during defined stages across the life span—the perinatal and pubertal periods. A large body of work has defined how sex hormones and, in particular, androgens exert masculinizing effects on brain regions, which then trigger a cascade of sexually dimorphic downstream effects throughout the brain, as mediated through sex hormone receptor distribution (Bale & Epperson, 2015). Taken together, these processes provide a foundation for ways in which sex can affect neurodevelopment and lateralization.

SEX CHROMOSOME GENETICS AND NEURODEVELOPMENT

In comparison to hormone effects, genetic mechanisms underlying sex differences have not been studied as much, despite evidence that they have a direct influence on neurodevelopment (Carruth, Reisert, & Arnold, 2002). This is surprising given the substantial biological differences between males and females in genetic makeup, primarily driven by the unique nature of sex chromosomes, in which typically developing females carry two copies of the X chromosome, whereas typically developing males have one X and one Y chromosome. The difference in sex chromosome karyotype is also associated with a significant differential in the distribution of genetic material. Indeed, the X chromosome is three times as large as the Y chromosome and carries ~7 times as many coding genes. This lack of homology between the sex chromosomes is somewhat mitigated by a process of X-inactivation in females in which one of the X chromosomes in each cell is randomly inactivated. Regions on the X chromosome still escape inactivation, however, including loci that are homologous to regions on the Y chromosome (pseudoautosomal regions) as well as escapee genes that preferentially bypass the inactivation process (see Figure 9.2).

These gene regions represent potential sources of genetic discrepancies between males and females. These loci may result in varying levels of gene expression correlated with sex chromosome complement. Alternatively, they may provide rescue mechanisms for aberrant genes through the presence of additional copies of sex chromosome genes involved in neurodevelopmental pathways, providing a further potential link to similar processes in male-predominant disorders, such as developmental dyslexia.

SEX CHROMOSOME ANEUPLOIDIES AS MODEL SYSTEMS FOR NEURODEVELOPMENTAL DISORDERS

Discriminating effects of hormonal and genetic influence in humans remains a central challenge, however, given that these processes often interact and exert their influence during similar periods of development, thus making it difficult to separate these dimensions. Human disease models may provide a potential means for better understanding these effects, namely SCAs—conditions in which individuals have an abnormal number

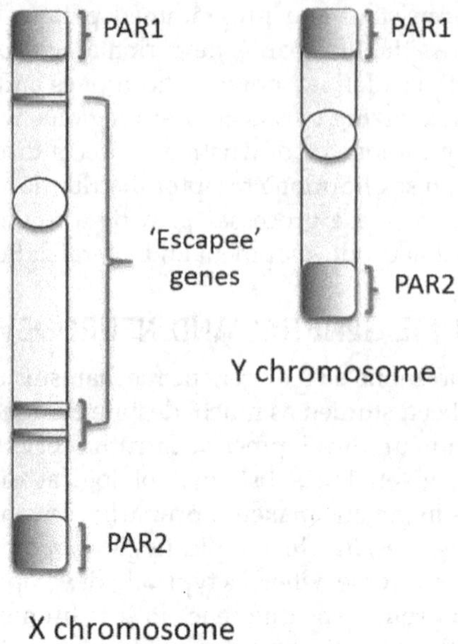

Figure 9.2. The sex chromosomes comprise the X and Y chromosomes, which differ substantially in size and the number of genes. Although one of the two X chromosomes in 46XX females are inactivated, distal ends of the X chromosome (PAR1 and PAR2), which are homologous to regions on the Y chromosome, are not inactivated. Similarly, certain genes on the X chromosome escape the X-inactivation process.

of sex chromosomes. By definition, individuals with SCAs have varying degrees of genetic material from the sex chromosomes, which essentially represent the gain or loss of an entire chromosome. Turner syndrome, in which females have one less chromosome (45X karyotype), and Klinefelter syndrome, which is characterized by an additional X chromosome in males (47XXY karyotype) are two of the most prominent conditions. Both of these SCAs have consistently been characterized with unique cognitive profiles. Although individuals with TS and KS have overall IQs within the average range, subcomponents of IQ indicate a specific discrepancy between verbal and nonverbal domains (Hong & Reiss, 2014). Females with TS typically demonstrate intact verbal IQ scores in the average range in contrast to decreased scores in nonverbal domains, which encompass visuospatial, arithmetic, and executive functions (Hong, Scaletta Kent, & Kesler, 2009). Yet, males with KS tend to have intact nonverbal scores and decreases in verbal IQ (Boada, Janusz, Hutaff-Lee, & Tartaglia, 2009). Not surprisingly, measurements in these cognitive domains are mirrored by correlated difficulties in academic domains that draw on these functions, with girls with TS demonstrating increased prevalence of mathematics learning disability up to 50% (Mazzocco, 2001) and boys with KS having much higher rates of language-based disorders in the range of 50%–75% (Bender, Puck,

Salbenblatt, & Robinson, 1986; Graham, Bashir, Stark, Silbert, & Walzer, 1988). Individuals with SCAs are also much more likely to utilize specialized education services compared with peers, further demonstrating the potential of SCAs to act as model conditions to investigate sex-related features of LDs. Interestingly, SCAs have also been found to occur at much higher rates in youth with developmental language disorders (between 3%–5%) compared with the general population (Robinson, 1991; Simpson et al., 2014).

Dose-dependent Relationship in Sex Chromosome Aneuploidies

It is particularly interesting to note the relationship between individuals with TS and KS and typically developing controls because there appears to be an inverse pattern in the cognitive domains affected (see Figure 9.3). Mean scores on IQ subdomains are grossly correlated with sex chromosome number, suggesting a dose-dependent relationship such that a relative decrease in sex chromosome number (and associated gene expression) is associated with impairments in visuospatial function, whereas a relative increase is associated with language-based difficulties (Hong & Reiss, 2014).

A similar dose-dependent pattern is reflected in structural neuroanatomy in which the presence of two rather than one X chromosome results in increased volume in the parieto-occipital regions, as would be the case

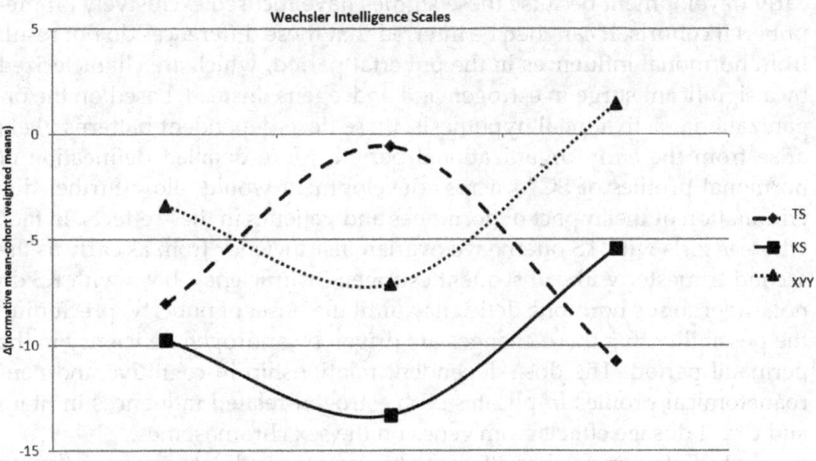

Figure 9.3. Sex chromosome aneuploidies have characteristic neurocognitive profiles. Interestingly, these profiles appear to follow dose-dependent patterns for intelligence domains. Although general intelligence (FSIQ) falls within the normal range, the absence of an X chromosome (TS) is associated with intact verbal ability (VIQ) and deficits in noverbal reasoning (PIQ), whereas the opposite trend holds true for cohorts with an additional sex chromosome (KS and XYY groups).

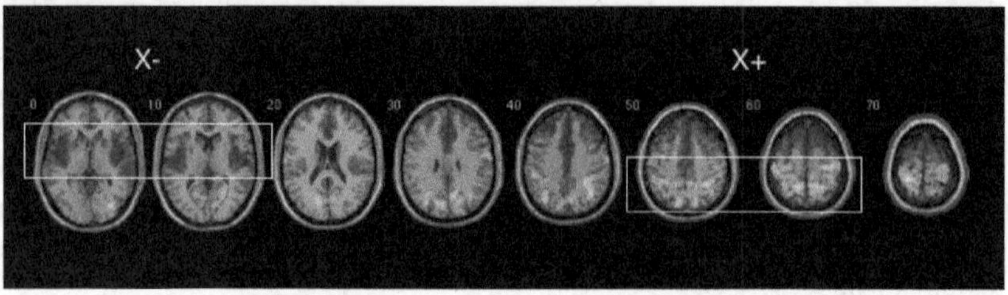

Figure 9.4. Neuroanatomical comparisons of gray matter volume between individuals with Turner and Klinefelter syndromes find that decreased X chromosome complement (X–) is associated with increased volume in the temporo-insular regions, whereas increased X chromosome complement (X+) is associated with increased volume in parieto-occipital structures.

in typically developing girls compared with girls with TS (XX *vs.* X) and in boys with KS compared with control males (XXY *vs.* XY). Similarly, in the inverse direction, the relative absence of an X chromosome is associated with increases in gray matter volume in temporo-insular regions (see Figure 9.4) (Hong et al., 2014).

These findings have been found across development to varying degrees in children and adults with SCAs and in other polyploidy SCAs such as XXX and XYY syndromes, suggesting that it may be tied to sex chromosome number overall rather than the X chromosome alone (Lentini, Kasahara, Arver, & Savic, 2013; Raznahan et al., 2016). It is important to consider that these cognitive and neuroanatomical findings are already present in early development because these studies have focused exclusively on prepubertal cohorts. It can then be inferred that these differences do not result from hormonal influences in the pubertal period, which are characterized by a significant surge in estrogen and androgens. Instead, based on the organizational-activational hypothesis, these dose-dependent patterns likely arise from the early organizational period. More detailed delineation of hormonal profiles of SCAs across development would allow further discrimination of the impact of hormones and genetics in these effects. In fact, whereas girls with TS often have ovarian insufficiency from as early as the second trimester with subsequent estrogen insufficiency, boys with KS do not undergo sex hormone deficiency until the onset of puberty, precluding the possibility that these changes are driven by androgen deficiency in the perinatal period. The dose-dependent relationship in cognitive and neuroanatomical profiles implicates both estrogen-related influences in utero and direct dosage effects from genes on the sex chromosomes.

In fact, prior work has attempted to isolate candidate genes on the sex chromosome that may be tied to cognitive functions, including findings that the cognitive profile associated with TS may be associated with genes on the short arm of the X chromosome (Ross, Roeltgen, Kushner, Wei, & Zinn, 2000). Remarkably, investigation of gene expression patterns in this same locus in males with KS demonstrated that overexpression of genes

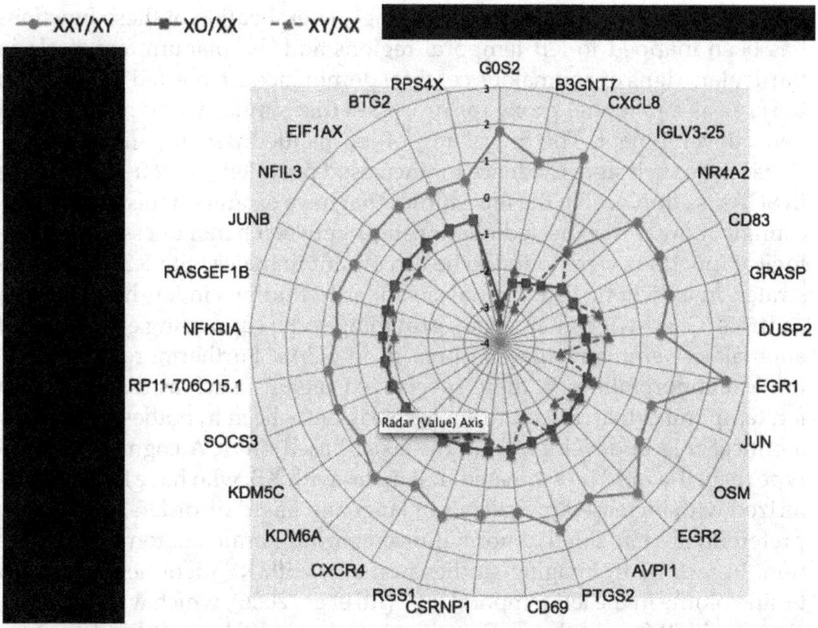

Figure 9.5. Comparison of RNA-Seq gene expression profiles between cohorts with a relative difference in X dosage (TS *vs.* XX, XY *vs.* XX, and XY vs. XXY), identify a number of genes across the sex chromosomes and autosomes with dose-dependent expression levels.

in this same region is associated with the verbal cognition deficits (Vawter, Harvey, & DeLisi, 2007). Conclusive identification of single genes on the sex chromosomes that fully explain the observed cognitive profiles in SCAs is still lacking. An alternate hypothesis is that sex chromosomes exert genetic influences across the entire network of gene expression, including on the autosomes, mediated through downstream effects and gene–gene interactions. In fact, recent work supports the presence of dose-dependent relationships in gene expression profiles across SCAs and typical controls (see Figure 9.5). It is also possible that genetic effects are directly correlated with hormonal influence given the location of the human androgen receptor (AR) gene on the X chromosome. Although the AR gene is subject to X-inactivation, mechanisms through which AR polymorphism mediates testosterone sensitivity in SCAs warrants further study.

SYNTHESIZING FINDINGS IN SEX CHROMOSOME ANEUPLOIDIES WITH THE GESCHWIND-GALABURDA HYPOTHESIS AND LANGUAGE DEVELOPMENT

The specialization of neural anatomy and related function via highly programmed patterns of cerebral lateralization is one of the central tenets of the GGH. Specifically regarding language, lateralization of these functions has been mapped to left temporal regions and the planum temporale in particular. Signs of anomalous cerebral dominance are posited by the GGH to serve as a potential proxy for processes that similarly underlie aberrant neurodevelopment. The prevalence of left-handedness in patient populations is one such sign. Evidence for increased prevalence of left-handedness in SCAs is equivocal in the few studies that have examined this issue. Interestingly, however, expected relationships between handedness and neurological functions were found to be discordant in males with KS, as demonstrated in a dichotic listening task, similar to findings in left-handed male controls (Geschwind et al., 1998), providing some supporting evidence that anomalous symmetry may be present in SCAs. Furthermore, neuroanatomical abnormalities in these specialized regions, such as disruptions in left temporal cytoarchitecture, have consistently been hypothesized to play a critical role in developmental dyslexia. Based on SCA cognitive phenotype then, it would be estimated that males with KS, who have been characterized with increased prevalence of language-based disorders, might have preferentially left-sided abnormalities regarding brain anatomy and function. In fact, neuroimaging studies in males with KS identified decreased brain volume in the left temporal lobe (Itti et al., 2006), which would be consistent with this hypothesis. Functional studies in KS have utilized positron emission tomography and found evidence for increased right-hemisphere cerebral blood flow in parietal, temporal, and prefrontal motor areas in men with KS, which correlated with the degree of verbal cognitive deficits (Itti et al., 2003). Although the localization of increased utilization in right-sided structures may initially appear counterintuitive, it is compatible with well-established compensatory mechanisms of function in which contralateral brain regions demonstrate relative increases in response to deficits in homologous anatomy in the opposite hemisphere. Yet, verbal skills are typically preserved in females with TS, which would be putatively associated with intact leftward temporal anatomy. In contrast, the prominent visuospatial difficulties observed in TS would be consistent with a right-hemisphere deficit based on the lateralization hypothesis. Again, although data are sparse, existing neuroimaging studies found that adult women with TS had increased patterns of asymmetry in the superior temporal sulci on par with that of typically developing adult males, in contrast to typically developing females who demonstrated more symmetrical neuroanatomy (Leroy et al., 2015), though the relationship of this anatomical finding with language ability was not investigated.

A UNIFIED HYPOTHESIS RELATING SEX CHROMOSOMES TO DEVELOPMENTAL DYSLEXIA

There is insufficient evidence to construct a parsimonious model fully integrating sex chromosome function with developmental dyslexia. There remains a compelling basis that a relationship between sex chromosomes and LDs exists, however, not the least of which is the significant overlap in cognitive profiles in a distinct dose-dependent fashion on learning phenotypes, as well as the partial overlap in neuroanatomical regions demonstrating aberrant structure in SCAs and developmental dyslexia, specifically the superior temporal gyrus/sulcus and the supramarginal gyrus. There is also the possibility that there may be convergent genetic pathways that act in direct or pleiotropic mechanisms between the sex chromosomes and candidate genes for developmental dyslexia to affect brain patterning and function. Finally, although some evidence exists that sex chromosomes may directly affect brain symmetry and its related functions, further study is needed to assess the nature of this relationship to the pathophysiology of LDs. This raises the larger conceptual question of the function of sex in the larger model of understanding developmental dyslexia. One might consider that processes leading to LDs are directly mediated through sex-specific pathways, as partially demonstrated through the documented role of sex chromosomes and hormones in lateralization and LD prevalence. The relative paucity of research finding distinct LD phenotypes between males and females, however, suggests that mediation mechanisms alone would not adequately account for observations in LDs, giving rise to the possibility that sex-specific factors play a moderating role in the relationship between pathophysiological mechanisms of LDs and observed phenotype. In sum, there is mounting evidence that sex plays a central role in neurodevelopment, cognitive phenotype, and aberrant trajectories of learning, such as in LDs. It is also clear that the mechanisms by which sex influences these processes are complex. Further study is needed to examine these relationships in future research, with the ultimate goal of improving our understanding of LDs and providing tangible targets of intervention for individuals affected with these conditions.

REFERENCES

Bale, T. L., & Epperson, C. N. (2015). Sex differences and stress across the lifespan. *Nature Neuroscience, 18*(10), 1413–1420. doi:10.1038/nn.4112

Bender, B. G., Puck, M. H., Salbenblatt, J. A., & Robinson, A. (1986). Dyslexia in 47,XXY boys identified at birth. *Behavior Genetics, 16*(3), 343–354.

Boada, R., Janusz, J., Hutaff-Lee, C., & Tartaglia, N. (2009). The cognitive phenotype in Klinefelter syndrome: A review of the literature including genetic and hormonal factors. *Developmental Disabilities Research Review, 15*(4), 284–294. doi:10.1002/ddrr.83

Carruth, L. L., Reisert, I., & Arnold, A. P. (2002). Sex chromosome genes directly affect brain sexual differentiation. *Nature Neuroscience, 5*(10), 933–934. doi:10.1038/nn922

Geschwind, D. H., Gregg, J., Boone, K., Karrim, J., Pawlikowska-Haddal, A., Rao, E.,… Nelson, S. F. (1998). Klinefelter's syndrome as a model of anomalous cerebral laterality: Testing gene dosage in the X chromosome pseudoautosomal region using a DNA microarray. *Developmental Genetics, 23*(3), 215–229. doi:10.1002/(SICI)1520-6408(1998)23:3<215::AID-DVG7>3.0.CO;2-X

Geschwind, N., & Galaburda, A. M. (1985a). Cerebral lateralization. Biological mechanisms, associations, and pathology: I. A hypothesis and a program for research. *Archives of Neurology, 42*(5), 428–459.

Geschwind, N., & Galaburda, A. M. (1985b). Cerebral lateralization. Biological mechanisms, associations, and pathology: II. A hypothesis and a program for research. *Archives of Neurology, 42*(6), 521–552.

Geschwind, N., & Galaburda, A. M. (1985c). Cerebral lateralization. Biological mechanisms, associations, and pathology: III. A hypothesis and a program for research. *Archives of Neurology, 42*(7), 634–654.

Graham, J. M., Jr., Bashir, A. S., Stark, R. E., Silbert, A., & Walzer, S. (1988). Oral and written language abilities of XXY boys: Implications for anticipatory guidance. *Pediatrics, 81*(6), 795–806.

Hawke, J. L., Olson, R. K., Willcut, E. G., Wadsworth, S. J., & DeFries, J. C. (2009). Gender ratios for reading difficulties. *Dyslexia, 15*(3), 239–242. doi:10.1002/dys.389

Hawke, J. L., Wadsworth, S. J., Olson, R. K., & DeFries, J. C. (2007). Etiology of reading difficulties as a function of gender and severity. *Reading and Writing, 20*(1-2), 13–25. doi:10.1007/s11145-006-9016-z

Hedges, L. V., & Nowell, A. (1995). Sex differences in mental test scores, variability, and numbers of high-scoring individuals. *Science, 269*(5220), 41–45.

Hong, D. S., Hoeft, F., Marzelli, M. J., Lepage, J. F., Roeltgen, D., Ross, J., & Reiss, A. L. (2014). Influence of the X-chromosome on neuroanatomy: Evidence from Turner and Klinefelter syndromes. *Journal of Neuroscience, 34*(10), 3509–3516. doi:10.1523/JNEUROSCI.2790-13.2014

Hong, D. S., & Reiss, A. L. (2014). Cognitive and neurological aspects of sex chromosome aneuploidies. *Lancet Neurology, 13*(3), 306–318. doi:10.1016/S1474-4422(13)70302-8

Hong, D., Scaletta Kent, J., & Kesler, S. (2009). Cognitive profile of Turner syndrome. *Developmental Disabilities Research Review, 15*(4), 270–278. doi:10.1002/ddrr.79

Hyde, J. S. (2016). Sex and cognition: Gender and cognitive functions. *Current Opinion in Neurobiology, 38*, 53–56. doi:10.1016/j.conb.2016.02.007

Itti, E., Gaw Gonzalo, I. T., Boone, K. B., Geschwind, D. H., Berman, N., Pawlikowska-Haddal, A.,…Swerdloff, R. S. (2003). Functional neuroimaging provides evidence of anomalous cerebral laterality in adults with Klinefelter's syndrome. *Annals of Neurology, 54*(5), 669–673. doi:10.1002/ana.10735

Itti, E., Gaw Gonzalo, I. T., Pawlikowska-Haddal, A., Boone, K. B., Mlikotic, A., Itti, L.,… Swerdloff, R. S. (2006). The structural brain correlates of cognitive deficits in adults with Klinefelter's syndrome. *Journal of Clinical Endocrinology and Metabolism, 91*(4), 1423–1427. doi:10.1210/jc.2005-1596

Lentini, E., Kasahara, M., Arver, S., & Savic, I. (2013). Sex differences in the human brain and the impact of sex chromosomes and sex hormones. *Cerebral Cortex, 23*(10), 2322–2336. doi:10.1093/cercor/bhs222

Leroy, F., Cai, Q., Bogart, S. L., Dubois, J., Coulon, O., Monzalvo, K.,…Dehaene-Lambertz, G. (2015). New human-specific brain landmark: The depth asymmetry of superior temporal sulcus. *Proceedings of the National Academy of Sciences USA, 112*(4), 1208–1213. doi:10.1073/pnas.1412389112

Mazzocco, M. M. (2001). Math learning disability and math LD subtypes: Evidence from studies of Turner syndrome, fragile X syndrome, and neurofibromatosis type 1. *Journal of Learning Disabilities, 34*(6), 520–533.

Moll, K., Kunze, S., Neuhoff, N., Bruder, J., & Schulte-Korne, G. (2014). Specific learning disorder: Prevalence and gender differences. *PLoS One, 9*(7), e103537. doi:10.1371/journal.pone.0103537

Ogle, L., Sen, A., Pahlke, E., Jocelyn, L., Kostberg, D., Roey, S., & Williams, T. (2003). *International comparisons in fourth-grade reading literacy: Findings from the Progress in International Literacy Study (PIRLS) of 2001* (NCES 2003-073). Washington, DC: U.S. Government Printing Office.

Olson, R. K. (2002). Dyslexia: Nature and nurture. *Dyslexia, 8*(3), 143–159. doi:10.1002/dys.228

Raznahan, A., Lee, N. R., Greenstein, D., Wallace, G. L., Blumenthal, J. D., Clasen, L. S., & Giedd, J. N. (2016). Globally divergent but locally convergent X- and Y-chromosome influences on cortical development. *Cerebral Cortex, 26*(1), 70–79. doi:10.1093/cercor/bhu174

Robinson, R. J. (1991). Causes and associations of severe and persistent specific speech and language disorders in children. *Developmental Medicine and Child Neurology, 33*(11), 943–962.

Ross, J. L., Roeltgen, D., Kushner, H., Wei, F., & Zinn, A. R. (2000). The Turner syndrome-associated neurocognitive phenotype maps to distal Xp. *American Journal of Human Genetics, 67*(3), 672–681. doi:10.1086/303039

Rutter, M., Caspi, A., Fergusson, D., Horwood, L. J., Goodman, R., Maughan, B.,…Carroll, J. (2004). Sex differences in developmental reading disability: New findings from 4 epidemiological studies. *The Journal of the American Medical Association, 291*(16), 2007–2012. doi:10.1001/jama.291.16.2007

Shaywitz, S. E., Shaywitz, B. A., Fletcher, J. M., & Escobar, M. D. (1990). Prevalence of reading disability in boys and girls: Results of the Connecticut Longitudinal Study. *The Journal of the American Medical Association, 264*(8), 998–1002.

Simpson, N. H., Addis, L., Brandler, W. M., Slonims, V., Clark, A., Watson, J.,…Consortium, SLI. (2014). Increased prevalence of sex chromosome aneuploidies in specific language impairment and dyslexia. *Developmental Medicine and Child Neurology, 56*(4), 346–353. doi:10.1111/dmcn.12294

Stoet, G., & Geary, D. C. (2013). Sex differences in mathematics and reading achievement are inversely related: Within- and across-nation assessment of 10 years of PISA data. *PLoS One, 8*(3), e57988. doi:10.1371/journal.pone.0057988

Vawter, M. P., Harvey, P. D., & DeLisi, L. E. (2007). Dysregulation of X-linked gene expression in Klinefelter's syndrome and association with verbal cognition. *American Journal of Medical Genetics B Neuropsychiatric Genetics, 144B*(6), 728–734. doi:10.1002/ajmg.b.30454

Voyer, D., Voyer, S., & Bryden, M. P. (1995). Magnitude of sex differences in spatial abilities: A meta-analysis and consideration of critical variables. *Psychology Bulletin, 117*(2), 250–270.

Willingham, W. W., & Cole, N. S. (1997). *Gender and fair assessment.* Mahwah, NJ: Lawrence Erlbaum Associates.

Wolraich, M. L., McKeown, R. E., Visser, S. N., Bard, D., Cuffe, S., Neas, B.,…Danielson, M. (2014). The prevalence of ADHD: Its diagnosis and treatment in four school districts across two states. *Journal of Attention Disorders, 18*(7), 563–575. doi:10.1177/1087054712453169

CHAPTER 10

Animal Models of Early Neural Disruption

Sex Differences, Neuroplasticity, and Implications for Dyslexia

R. Holly Fitch, Courtney Hill Bodge, Caitlin Szalkowski, and Amanda Smith

SUMMARY

This chapter highlights work on sex differences in atypical cognitive outcomes, particularly those relevant to language and reading outcomes, as observed in induced-injury and genetic animal models. These findings can be brought to bear on original ideas put forward by Geschwind and Galaburda (1985a, b, c), specifically concerning why more males are diagnosed with dyslexia. Although many questions remain, data clearly show that some factor(s)—almost certainly biological in origin—account for a heightened risk for neurodevelopmental anomalies that underlie cognitive impairments (e.g., dyslexia) in males.

INTRODUCTION

A major premise of the original Geschwind-Galaburda Hypothesis (GGH; 1985a, b, c; see Chapter 1) was that some fetal/neonatal factor leads the male brain to developmentally differ from the female brain. This factor could relate to the disproportionate incidence of developmental disabilities, such as dyslexia, as well as left-handedness, among males. Scant research has focused on the etiology of human handedness—possibly due to a shortage of useful animal models or the unclear epidemiologic significance of handedness in an era when translational potential for improvement in human health is crucial to research funding success (but see Annett, 2002; Corballis, 2003; Häberling, Corballis, & Corballis, 2016). Similarly, scant research in the past three decades has explored sex differences in human neurodevelopment, or cognitive/language outcomes associated with neurodevelopmental disruption (e.g., genetic and other risk factors associated with dyslexia). This dearth of research may reflect a tendency to exclusively use males in research, based on simplicity (humans) or cost (animals), as well as a political aversion to evidence of biologic sex differences in an era of heightened awareness about diversity and equality (Becker et al., 2005; McCarthy, Arnold, Ball, Blaustein, & De Vries, 2012; McCullough et al., 2014). Notably, the National Institutes of Health (NIH) has formally recognized the importance of sex and hormones in modulating disease and pharmaceutical efficacy, implementing a call for renewed emphasis on inclusion of females in research (see NIH policy NOT-OD-15-102). As a result, we are beginning to see an improved understanding of sex differences in adult disorders such as stroke, as well as the role of hormonal status in influencing adult post-injury clinical outcomes (for review, see Ahnstedt, McCullough, & Cipolla, 2016). Yet surprisingly little research has been undertaken to address pre- and postnatal and childhood sex differences, including factor(s) underlying the dramatic sex difference in the prevalence of neurodevelopmental disorders (NDs).

Although the current literature provides limited evidence congruent with the GGH position that sex differences in risk center on a perinatal delay in left-hemisphere development in males (with a corresponding extension of the early window of neural vulnerability), evidence does validate a significant discrepancy in deleterious perinatal events, including more spontaneous abortions, placental abnormalities, birth complications, and premature deliveries among male infants, as well as worse outcomes for males following these high-risk events (Gualtieri & Hicks, 1985). Findings also support a significant clinical discrepancy in developmental disorders, ranging from autism spectrum disorder (ASD; ~4:1) to attention-deficit/hyperactivity disorder (~5-7:1) to cerebral palsy (~1.2:1) to dyslexia (~2-3:1) (Polyak, Rosenfeld, & Girirajan, 2015, Taylor et al., 2016). Unfortunately, the smaller numbers of females affected by NDs has led some researchers to focus their research on affected males, or pool across sex, based on insufficient numbers to experimentally evaluate sex as a variable. Overall, the GGH notion that some early factor may alter the development and organization of the male versus female brain and, thus, modulate sex differences in outcome following disruptive neurodevelopmental events, remains a very promising model that could help explain the sometimes dramatic sex differences in disorder incidence and early brain injury outcomes. This chapter aims to review and integrate some of these findings, building on the premises put forward by the GGH decades ago.

SEX DIFFERENCES IN ANIMAL MODELS OF EARLY BRAIN INJURY AND ASSOCIATED NEURODEVELOPMENTAL DISORDERS

The cascading events of early sexual differentiation—beginning with transcription in males of the Sry gene (on the Y chromosome), regression/enhancement of primordial sex-specific internal structures, followed by development of the testes/ovaries and prenatal secretion of sex-specific hormones (especially high levels of androgens in males)—is described in detail in de Vries (see Chapter 4; see also Hines, 2008; Knickmeyer & Baron-Cohen, 2006; McCarthy, 2008). The enhanced prenatal exposure to testosterone could exacerbate the incidence of NDs, leading to male bias as well as more brothers among probands affected (James, 2008). The perspective of masculinizing processes as a risk factor coincides with the extreme male brain

(EMB) theory of ASD (Baron-Cohen et al., 2011; Wen & Wen, 2014). Unfortunately, animal studies of early genetic variables and subsequent cognitive outcomes that might provide physiologic insight to the EMB theory typically overlook the important role of sex and sex hormones. That is, most phenotyping studies of genetically manipulated animals (e.g., knockout mice) either use males exclusively or pool across sex without reporting sex-specific outcomes. Although there are no studies using ASD animal models to assess the EMB theory of which we are aware, some evidence does indicate that neonatal estrogen exposure can reverse perseverative behaviors (a core ASD-like feature) in at least one mouse ASD model (Macrì et al., 2010).

Overall evidence substantiates a more negative outcome for males following deleterious perinatal events, including a more robust decrement in IQ associated with premature delivery (Smith, Alexander, Rosenkrantz, & Fitch, 2014) as well as specific perinatal events such as intracranial hemorrhage or stroke (Fuentes, Deotto, Desrocher, deVeber, & Westmacott, 2016; Raz, Lauterbach, Hopkins, Glogowski et al., 1995). Findings also support the original GGH hypothesis that immune factors could play a role in these differences, given an elevated prevalence of children with NDs among mothers with systemic lupus (Vinet et al., 2014) as well as evidence that estrogen may protect against deleterious effects of neonatal inflammation (Pansiot et al., 2016; see also McCarthy, 2008). Thus, it is possible that perinatal exposure to inflammatory factors (typically elevated in conjunction with brain injury) could influence deleterious outcomes and increase risk by interacting with hormonal modulation of neural injury, cell death mechanisms, or even sex-specific post-injury mechanisms of reorganization.

Animal research has built on an extensive foundation of developmental sex-difference research by early pioneers (see McCarthy et al., 2012 for review) to more recently examine the developmental role of sex hormones in modulating early brain injury outcomes (Hill, Alexander, McCullough, & Fitch, 2011; Hill & Fitch, 2012; Hill, Threlkeld, & Fitch 2011; Lang & McCullough, 2008; Nuñez & McCarthy, 2008; Smith et al., 2014; Smith, Rosenkrantz, & Fitch, 2016; Waddell, Hanscom, Shalon Edwards, McKenna, & McCarthy, 2016; Zup, Edwards, & McCarthy, 2014). In general, these reports and others collectively reaffirm female protection, or alternately, an exacerbation of deleterious neonatal outcomes in males. For example, substantially elevated levels of testosterone present in human male fetuses during gestation and early postnatal life may enhance neuronal excitotoxicity following insult (Du et al., 2004; Lang & McCullough, 2008; Nuñez & McCarthy, 2008; Yang et al., 2002; Zup et al., 2014). This enhanced excitotoxicity could, in turn, contribute to an increase in cognitive deficits among males.

We assessed an induced-microgyria rat model that exhibits cortical malformations as seen in human postmortem dyslexic brains to test this model as it may apply to reading-related outcomes (Fitch, Alexander, & Threlkeld, 2013; Galaburda , Sherman, Rosen, Aboitiz, & Geschwind, 1985).

Results with this model showed that cortical malformations as seen in dyslexia were, in fact, associated with deficits in the encoding and discrimination of rapidly changing acoustic stimuli. These processing defects could relate to disruptions in the developmental acquisition of speech, language, and ultimately reading in humans (Clark, Rosen, Tallal, & Fitch, 2000; Fitch & Tallal, 2003; Fitch, Tallal, Brown, Galaburda, & Rosen, 1994). Because our original results were specific to male microgyric rats, we replicated our assessments in microgyric male and female rats, as well as male and female mice with other types of cortical developmental malformations (ectopias). Surprisingly, we consistently found rapid acoustic processing deficits present only in males (Fitch, Brown, Tallal, & Rosen, 1997; Peiffer, Rosen, & Fitch, 2002, 2004). Moreover, we found a shift in medial geniculate nucleus (auditory thalamic nucleus) cell size in male rats with induced cortical microgyria, as had previously been seen in postmortem human dyslexic brains (Galaburda, Menard, & Rosen, 1994). These effects were not seen in intact female rats with microgyria (who also failed to show behavioral deficits), but, interestingly, they were seen in female rats that had been neonatally treated with testosterone propionate (Herman, Galaburda, Fitch, Carter, & Rosen, 1997). We subsequently reported similar deficits on a rapid acoustic processing task among male rats with induced neonatal injury typical of that seen in preterm infants (and who are, in turn, at risk for subsequent language and reading disorders) (McClure, Threlkeld, Rosen, & Fitch, 2006). More important, these same deficits were not significant in intact female rats with comparable hypoxic-ischemic (HI) brain injuries, although deficits were significant in female rats if they were treated neonatally with testosterone propionate (Hill, Threlkeld, & Fitch, 2011).

In interpreting our results, we were interested in whether neonatal males and females might diverge in the proportional activation of the caspase-dependent and -independent pathways that lead to apoptotic death and tissue loss following brain injury (Lang & McCullough, 2008; Renolleau, Fau, & Charriaut-Marlangue, 2008; Zhu et al., 2006). We hypothesized that this difference may play a role in the heightened vulnerability to poor language and reading outcomes following perinatal brain damage among males (Hagberg et al., 2004; Nijboer et al., 2007; Renolleau et al., 2007). Prior evidence has shown that both Parp-1 and AIF (biproducts of one apoptotic cell death pathway) are elevated in the brains of male mice following postnatal hypoxic-ischemic injury and are higher than levels seen in comparably injured females (Zhu et al., 2006). Significant protection from postnatal injury has also been shown in male Parp-1 knockout mice but not in females (Hagberg et al., 2004). Others reported that cytochrome-c and various caspases (active in the caspase-dependent cell death pathway) are elevated in female as compared with male mice following neonatal brain injury, and the inhibition of caspase cleavage is neuroprotective in female (but not male) animals following neonatal injury (Hagberg et al., 2004; Nijboer et al., 2007; Renolleau et al., 2007). These findings parallel similar sex differences observed in adult stroke animal models (McCullough & Hurn, 2003). In fact, we found that manipulations of the apoptotic cascade focused on an X-linked inhibitor of apoptosis selectively enhanced deleterious outcomes in female rats with neonatal brain injury (Hill et al., 2011).

It appears, however, that the sex difference in apoptotic response to neural injury is at least partly modulated in adults by the presence of ovarian estrogen in females (Carswell et al., 2001; McCullough & Hurn, 2003; Turtzo & McCullough, 2008). This is in part why adult pre-menopausal women have a lower rate of naturally occurring stroke and show less sensitivity than males to the damaging effects of focal or global ischemic injury (reviewed in Herson, Koerner, & Hurn, 2009). Moreover, in animal models with a condition known to be a stroke risk factor (e.g., hypertension), females show less tissue damage than males following induced stroke (Liu et al., 2009). This female advantage has also been attributed, in part, to protective effects of ovarian steroid hormones because interventions that reduce estrogens (i.e., ovariectomy, estrogen receptor blockade, natural aging) act to eliminate sex differences in adult stroke outcome (McCullough & Hurn, 2003). The potential protective role of estrogen in response to brain injury in neonates remains questionable, however, because the ovaries do not produce significant quantities of estrogen during fetal and early postnatal life (see Fitch & Denenberg, 1998).

It is alternately possible that elevated exposure to androgens in males may influence sex differences in the activation of cell death pathways and, thus, alter sex-based cognitive and language outcomes following early brain injury. Paradoxical evidence, however, shows that neonatal androgen exposure in mouse models actually yields neuroprotective effects against stroke in adulthood (Persky et al., 2013).

SEX DIFFERENCES IN GENETIC MODELS OF NEURODEVELOPMENTAL DISORDERS

Studies of developmental sex differences in response to early injury, stress, inflammation, or toxin exposure cannot fully explain the significant sex differences in clinical ND prevalence. Although sex differences in the incidence and outcome from early neurologic events certainly account for some increase in intellectual developmental disabilities among males, other highly discrepant disorders such as ASD and dyslexia are most consistently associated with risk-gene mutations that alter development by disrupting the neurodevelopmental trajectory—not (as far as we know) through genetically modulated neural injury, inflammation, or injury-induced cell death.

In this regard, X-linked variants may be another obvious candidate for explaining gender bias in prevalence of NDs (because females inherit two X chromosomes, protecting them from recessive X-linked mutations). Yet, few ND association studies have actually found single-nucleotide variants, copy-number variants, or other mutations on the X-chromosome. Instead, most ND risk genes identified to date (e.g., ASD, dyslexia) are autosomal and, thus, seemingly equally likely to be inherited by either sex. Females may tolerate a higher load of gene mutations before reaching a clinical (behavioral) threshold for ND expression (Jaquemont et al., 2014), but a biologic mechanism for this putative threshold difference has not been identified. Some models further suggest that thresholds to respond to genetic liability for ND phenotypes may differ by sex as a function of the frequency and type of comorbid features, presence of disease-associated deletions and duplications, and/or family history of related phenotypes (Polyak et al., 2015; see also Chapter 14). Findings also show elevated maternally inherited mutations among randomly ascertained ND probands (Jaquemont et al., 2014). The implication of these results to the skewed incidence of NDs among males will require additional research before clear claims can be made.

Even with regard to sex differences in apoptotic response to brain injury previously discussed, it is apparent that genetic factors also play a role in these effects because embryonic neuronal cell cultures exposed to cytotoxic challenge still show differing pathways of cell death in the absence of circulating hormones (XY neurons predominantly utilizing the AIF-mediated caspase-independent apoptotic pathway and XX neurons activating the cytochrome-c, caspase-dependent apoptotic pathway) (Du et al., 2004). Moreover, females show other gene-linked advantages in response to injury, specifically via inhibitors of apoptosis (IAPs) such as X-linked IAP (XIAP; see Deveraux & Reed, 1999, for review). These endogenous inhibitors are known to affect the apoptotic pathway predominately activated in female neuronal death (Deveraux et al., 1998; Ferrer & Planas, 2003). Neonatal female rats treated with an inhibitor of XIAP (embelin) at the time of HI injury showed behavioral and acoustic processing deficits comparable to HI-injured male rats, although untreated females with HI injury did not (Hill et al., 2011).

Because neither hormonal nor X-linked genetic differences in apoptotic response to early injury appear to explain a sex difference in response to inherited or de novo gene mutations associated with the heightened incidence of NDs (e.g., dyslexia) in males, we decided to explore the possible interaction between sex and gene mutations associated with dyslexia by examining male and female rats treated embryonically with RNAi for *Dyx1c1*—the dyslexia-risk gene (rodent homolog). We predicted that female rats might show fewer (or no) deficits in acoustic discrimination compared with males. Results showed virtually identical patterns of auditory processing deficits in treated males and females, however, as well as comparable deficits in working memory associated with RNAi for *Dyx1c1* across both sexes (Szalkowski et al., 2011; Szalkowski, Booker, Truong, Rosen, & Fitch, 2013; Threlkeld et al., 2007). These results were puzzling, given prior consistency in findings of a protective female effect as measured on this same task using an injury model. We concluded that any female protective effect might be surpassed in the face of severe genetic disruption due to gene deactivation/mutation. This would be consistent with the fact that some females are affected by NDs, albeit at a lower incidence rate. Therefore, disruptions to the neural function of the *Dyx1c1* gene may be sufficiently potent to induce comparable deficits in both sexes. Alternately, females may show protection against only some forms of risk factors and mutations. The issue is difficult to tease apart using this animal model because it is not clear whether human females that were similarly affected by a *Dyx1c1* gene mutation would rise to clinical detection thresholds for dyslexia. Of course, this is a serious limitation of using animal models to study language-based NDs. Nonetheless, results did prove that females do not possess across-the-board protection against impaired outcomes in processing domains important to language and reading in all risk models.

Finally, epigenetics, such as imprinting, in which the same autosomal gene mutation may be differentially expressed depending on whether it is inherited from the mother or father (e.g., Prader-Willi or Angelman syndromes), must be considered. Similarly, X-linked imprinting may be relevant because boys always receive their X chromosome from their mother, whereas girls receive an X chromosome from both parents. Random X-deactivation, which is seen in females, could protect them from some deleterious X-imprinted effects on the maternal X. Moreover, because females produce their gametes early on in development (and these gametes include non-DNA cytoplasmic factors), whereas males generate gametes anew throughout life, the mechanisms of genetic transmission from parent to offspring may differ in some way that benefits females (see Chapter 14). These remain possible but unproven contributors to the observation that females show better outcomes from early neurodevelopmental disruption and, ultimately, are less likely to be diagnosed with NDs, including dyslexia.

CONCLUSION

Clearly, a wide range of developmental disruptions -- ranging from injury and/or inflammation that trigger cell death and tissue loss, to gene mutations that substantially alter the cascade of central nervous system development—can trigger plasticity in the developing brain. One evolutionary driver for plasticity could relate to the known tendency for neurodevelopmental processes to be highly canalized and protected by multiple back-up systems (Mitchell, 2015). Alternately, evolution may confer advantages on the ability to compensate for early injury, with neuroplasticity in the developing brain leading to beneficial outcomes that enhance survival and reproduction (e.g., recovery in children with early left hemispherectomy) (Skirrow et al., 2011). Early neuroplasticity could also simply reflect an epiphenomenon of the developmental process itself, leading to sometimes beneficial and sometimes profoundly maladaptive outcomes (Johnston, 2009; Kolb & Gibb, 2014).

The notion of indifferent or agnostic neuroplasticity (i.e., a process that cannot be labeled as ubiquitously good nor bad) has been advanced by Galaburda as a possible explanation for the brainwide anomalies observed in our joint research on rodent models of dyslexia-risk gene manipulations (Albert Galaburda, personal communication, June 2016). These anomalies include atypical cellular structure at the level of the brainstem, synaptic hyper-excitability, cortical neuronal migration anomalies, and clear behavioral deficits relevant to language and reading skills (e.g., acoustic processing, working memory) (reviewed in Fitch et al., 2013). The trigger threshold target (TTT) hypothesis supports this view, and has been advanced to explain sex differences in the prevalence of ASD. The TTT postulates that males have a lower threshold for plasticity during early neurodevelopment, and the resulting enhanced male neuroplasticity might lead to a higher prevalence of deleterious outcomes (Mottron et al., 2015). If true, this principle could explain both an elevated rate of poor outcomes in males following brain injury, as well as a heightened expression of clinical NDs in response to genetic risks and other ND risk factors (e.g., prenatal stress, early inflammation, toxin exposure). The authors, however, did not address the core explicit issue of what biologic mechanisms might lead males to show enhanced neurodevelopmental plasticity (or alternately, lower thresholds for maladaptive plasticity leading to ND diagnosis) relative to females.

In closing, the data reviewed suggest that an answer to the question of why more boys are diagnosed with dyslexia may lie in a complex interaction of genetic and hormonal differences that favor enhanced outcomes in females compared with males—at least when challenged by equal accumulations of mid- to moderate-level neurodevelopmental disruptions. Risks that exceed a threshold would still lead to deleterious outcomes in females, given that they do sometimes show poor outcomes (albeit it at a lower rate). The possibility that females who do surpass a risk threshold present with more severe symptoms than their male counterparts could reflect a tendency for females to remain subthreshold (and undiagnosed), with only serious accumulations of risk (e.g., injury, mutation, stress, toxins) characterizing females who rise to detection in randomly ascertained groups (e.g., Davies, 2014; Halladay et al., 2015; Mahone & Wodka, 2008; Rucklidge, 2010) (see Figure 10.1).

Although we still have much to learn about why males show a higher prevalence of NDs (including dyslexia), as well as worse outcomes in the cognitive, language, and reading domains following early brain injury or other neurodevelopmental risk factors, the theories proposed by Geschwind and Galaburda (1985a, b, c) survive to provide a contemporary model that still fits a wide array of cumulative data in the dyslexia field. The notion that the male brain may be more vulnerable to developmental challenge or disruption compared with the female brain continues to weave itself into modern and cutting-edge theories such as the TTT hypothesis (Mottron et al., 2015). Although the mechanisms originally postulated by

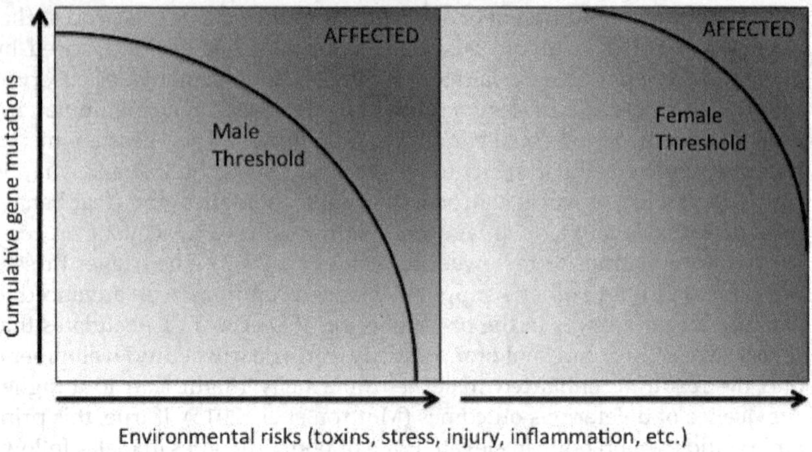

Figure 10.1. A schematic representation of net incidence levels for diagnosed neurodevelopmental disorders in the case of differential thresholds for phenotypic expression in males versus females.

the GGH may not turn out to be correct (i.e., male risk associated with delays in left-hemisphere development), the basic idea of differential vulnerability remains a viable basis for further exploration of the interactions of sex, hormones, genetics, and diverse early risk challenges (e.g., inflammation, stress, toxins, mutations) that modulate a complex trajectory of brain development, culminating in cognitive outcomes that encompass reading skills. Overall, the field of dyslexia research will dramatically benefit from an expanded focus on these variables, particularly with animal ND models that can provide experimental power not available to human clinical researchers (e.g., an ability to carefully assess core cognitive and sensory processing skills necessary in humans for language and reading proficiency) in the face of different early genetic and hormonal profiles. It is our hope that an increasing number of current studies will recognize and incorporate these profoundly integrative and prescient theories.

REFERENCES

Ahnstedt, H., McCullough, L. D., & Cipolla, M. J. (2016). The importance of considering sex differences in translational stroke research. *Translational Stroke Research, 7,* 261–273.

Annett, M. (2002). *Handedness and brain asymmetry: The right shift theory* (2nd ed.). East Sussex, UK: Psychology Press.

Baron-Cohen, S., Lombardo, M. V., Auyeung, B., Ashwin, E., Chakrabarti, B., & Knickmeyer, R. (2011). Why are autism spectrum conditions more prevalent in males? *PLoS Biology, 9,* e1001081.

Becker, J. B., Arnold, A. P., Berkley, K. J., Blaustein, J. D., Eckel, L. A., Hampson, E.… Young, E. (2005). Strategies and methods for research on sex differences in brain and behavior. *Endocrinology, 146,* 1650–1673.

Carswell, H. V., Anderson, N. H., Morton, J. J., McCulloch, J., Dominiczak, A. F, Macrae, I. M. (2001). Investigation of estrogen status and increased stroke sensitivity on cerebral blood flow after a focal ischemic insult. *Journal of Cerebral Blood Flow Metabolism, 20,* 931–936.

Clark, M., Rosen, G., Tallal, P., & Fitch, R. H. (2000). Impaired processing of complex auditory stimuli in rats with induced cerebrocortical microgyria. *Journal of Cognitive Neuroscience, 12*(5), 828–839.

Corballis, M. (2003). From mouth to hand: Gesture, speech, and the evolution of right-handedness. *Behavior and Brain Sciences, 26,* 199–208.

Davies, W. (2014). Sex differences in attention deficit hyperactivity disorder: Candidate genetic and endocrine mechanisms. *Frontiers in Neuroendocrinology, 35,* 331–346.

Deveraux, Q. L., & Reed, J. C. (1999). IAP family proteins: Suppressors of apoptosis. *Genes and Development, 13,* 239–252.

Deveraux, Q. L., Roy, N., Stennicke, H. R., Van Arsdale, T., Zhou, T., Srinivasula, S. M.,… Reed, J. C. (1998). IAPs block apoptotic events induced by caspase-8 and cytochrome c by direct inhibition of distinct caspases. *European Molecular Biology Organization Journal, 17,* 2215–2223.

Du, L., Bayir, H., Lai, Y., Zhang, X., Kochanek, P. M., Watkins, S. C., Graham, S. H., Clark, R. S. (2004). Innate gender-based proclivity in response to cytotoxicity and programmed cell death pathway. *Journal of Biological Chemistry,* 279(37), 38563–38570

Ferrer, I., & Planas, A. M. (2003). Signaling of cell death and cell survival following focal cerebral ischemia: Life and death struggle in the penumbra. *Journal of Neuropathology and Experimental Neurology, 62,* 329–339.

Fitch, R. H., Alexander, M., & Threlkeld, S. W. (2013). Early neural disruption and auditory processing outcomes in rodent models: Implications for developmental language disability. *Frontiers in Systems Neuroscience, Special Issue, 7,* 00058.

Fitch, R. H., Brown, C., Tallal, P., & Rosen, G. (1997). Effects of sex and MK-801 on auditory processing deficits associated with developmental microgyric lesions in rats. *Behavioral Neuroscience, 111,* 404–412.

Fitch, R. H., & Denenberg, V. H. (1998). A role for ovarian hormones in sexual differentiation of the brain. *Behavior and Brain Science, 21,* 311–352.

Fitch, R. H., & Tallal, P. (2003). Neural mechanisms of language based learning impairments: Insights from human populations and animal models. *Behavioral and Cognitive Neuroscience Reviews, 2*(3), 155–178.

Fitch, R. H., Tallal, P., Brown, C., Galaburda, A., & Rosen, G. D. (1994). Induced microgyria and auditory temporal processing in rats: A model for language impairment? *Cerebral Cortex, 4,* 260–270.

Fuentes, A., Deotto, A., Desrocher, M., deVeber, G., & Westmacott, R. (2016). Determinants of cognitive outcomes of perinatal and childhood stroke: A review. *Child Neuropsychology, 22,* 1–38.

Galaburda, A. M., Menard, M. T., & Rosen, G. D. (1994). Evidence for aberrant auditory anatomy in developmental dyslexia. *Proceedings of the National Academy of Sciences USA, 91,* 8010–8013.

Galaburda, A. M., Sherman, G. F., Rosen, G. D., Aboitiz, F., & Geschwind, N. (1985). Developmental dyslexia: Four consecutive cases with cortical anomalies. *Annals of Neurology, 18,* 222–233.

Geschwind, N., & Galaburda, A. M. (1985a). Cerebral lateralization. Biological mechanisms, associations, and pathology: I. A hypothesis and a program for research. *Archives of Neurology,* 42(5), 428–459.

Geschwind, N., & Galaburda, A. M. (1985b). Cerebral lateralization. Biological mechanisms, associations, and pathology: II. A hypothesis and a program for research. *Archives of Neurology,* 42(6), 521–552.

Geschwind, N., & Galaburda, A. M. (1985c). Cerebral lateralization. Biological mechanisms, associations, and pathology: III. A hypothesis and a program for research. *Archives of Neurology,* 42(7), 634–654.

Gualtieri, T., & Hicks, R. E. (1985). An immunoreactive theory of selective male affliction. *Behavioral and Brain Sciences, 8,* 427–441.

Häberling, I. S., Corballis, P. M., & Corballis, M. C. (2016). Language, gesture, and handedness: Evidence for independent lateralized networks. *Cortex, 82,* 72–85.

Hagberg, H., Wilson, M. A., Matsushita, H., Zhu, C., Lange, M., Gustavsson, M.,...Johnston, M. V. (2004). PARP-1 gene disruption in mice preferentially protects males from perinatal brain injury. *Journal of Neurochemistry, 90,* 1068–1075.

Halladay, A. K., Bishop, S., Constantino, J. N., Daniels, A. M., Koenig, K., Palmer, K.,... Szatmari, P. (2015). Sex and gender differences in autism spectrum disorder: Summarizing evidence gaps and identifying emerging areas of priority. *Molecular Autism, 6,* 36.

Herman, A. E., Galaburda, A. M., Fitch, R. H., Carter, A. R., & Rosen, G. D. (1997). Cerebral microgyria, thalamic cell size and auditory temporal processing in male and female rats. *Cerebral Cortex, 7,* 453–464.

Herson, P. S., Koerner, I. P., & Hurn, P. D. (2009). Sex, sex steroids, and brain injury. *Seminars in Reproductive Medicine, 27,* 229–239.

Hill, C. A., Alexander, M. L., McCullough, L. D., & Fitch, R. H. (2011). Inhibition of XIAP with embelin differentially affects behavioral outcome following neonatal hypoxia-ischemia in male and female rats. *Developmental Neuroscience, 33*(6), 494–504.

Hill, C. A., & Fitch, R. H. (2012). Sex differences in mechanisms and outcome of neonatal hypoxia ischemia in rodent models: Implications for sex-specific neuroprotection in clinical neonatal practice. *Neurology Research International* (special issue), Article ID 867531, 1–9. doi:10.1155/2012/867531.

Hill, C. A., Threlkeld, S. W., & Fitch, R. H. (2011). Early testosterone modulated sex differences in behavioral outcome following neonatal hypoxia ischemia in rats. *International Journal of Developmental Neuroscience, 29,* 381–388.

Hines, M. (2008). Early androgen influences on human neural and behavioural development. *Early Human Development, 84,* 805–807.

James, W.H. (2008). Further evidence that some male-based neurodevelopmental disorders are associated with high intrauterine testosterone concentrations. *Developmental Medicine and Child Neurology, 50,* 15–18.

Jaquemont, S., Coe, B. P., Hersch, M., Duyzend, M. H., Krumm, N., Bergmann, S.,... Eichler, E. E. (2014). A higher mutational burden in females supports a "female protective model" in neurodevelopmental disorders. *American Journal of Human Genetics, 94,* 415–425.

Johnston, M. V. (2009). Plasticity in the developing brain: Implications for rehabilitation. *Developmental Disabilities Research Review, 15,* 94–101.

Knickmeyer, R. C., & Baron-Cohen, S. (2006). Fetal testosterone and sex differences. *Early Human Development, 82,* 755–760.

Kolb, B., & Gibb, R. (2014). Searching for the principles of brain plasticity and behavior. *Cortex, 58,* 251–260.

Lang, J. T., & McCullough, L. D. (2008). Pathways to ischemic neuronal cell death: Are sex differences relevant? *Journal of Translational Medicine, 6,* 33. doi:10.1186/1479-5876-6-33

Liu, F., Li, Z., Li, J., Siegel, C., Yuan, R. & McCullough, L.D. (2009). Sex differences in caspase activation after experimental stroke. *Stroke, 40,* 1842–1848.

Macrì, S., Biamonte, F., Romano, E., Marino, R., Keller, F., & Laviola, G. (2010). Perseverative responding and neuroanatomical alterations in adult heterozygous reeler mice are mitigated by neonatal estrogen administration. *Psychoneuroendocrinology, 35,* 1374–1387.

Mahone, E. M., & Wodka, W. L. (2008). The neurological profile of girls with ADHD. *Developmental Disabilities Research Review, 14,* 276–284.

McCarthy, M. M. (2008). Estradiol and the developing brain. *Physiology Review, 88,* 91–124.

McCarthy, M. M., Arnold, A. P., Ball, G. F., Blaustein, J. D., & and de Vries, G. J. (2012). Sex differences in the brain: The not so inconvenient truth. *Journal of Neuroscience, 32,* 2241–2247.

McClure, M., Threlkeld, S., Rosen, G. & Fitch, R.H. 2006. Rapid auditory processing and learning deficits in rats with P1 versus P7 neonatal hypoxic-ischemic injury. *Behavioural Brain Research,* 172, 114–121.

McCullough, L. D., de Vries, G. J., Miller, V. M., Becker, J. B., Sandberg, K. & McCarthy, M. M. (2014). NIH initiative to balance sex of animals in preclinical studies: generative questions to guide policy, implementation, and metrics. *Biology of Sex Differences, 5,* 15.

McCullough, L. D., & Hurn, P. D. (2003). Estrogen and ischemic neuroprotection: An integrated view. *Trends in Endocrinology and Metabolism, 14,* 228–235.

Mitchell, K. (2015). *The genetics of neurodevelopmental disorders.* New York, NY: Wiley-Blackwell.

Mottron, L., Duret, P., Mueller, S., Moore, R. D., Forgeot d'Arc, B., Jacquemont, S., & Xiong, L. (2015). Sex differences in brain plasticity: A new hypothesis for sex ratio bias in autism. *Molecular Autism, 6,* 33.

Nijboer, C. H. A., Groenendaal, F., Kavelaars, A., Hagberg, H. H., van Bel, F., & Heijnen, C. J. (2007). Gender-specific neuroprotection by 2-iminobiotin after hypoxia-ischemia in the neonatal rat via a nitric oxide independent pathway. *Journal of Cerebral Blood Flow and Metabolism, 27*(2), 282–292.

Nuñez, J. L., & McCarthy, M. M. (2008). Androgens predispose males to GABAA-mediated excitotoxicity in the developing hippocampus. *Experimental Neurology, 210,* 699–708.

Pansiot, J., Pham, H., Dalous, J., Chevenne, D., Colella, M., Schwendimann, L.,…Baud O. (2016). Glial response to 17β-estradiol in neonatal rats with excitotoxic brain injury. *Experimental Neurology, 282,* 56–65.

Peiffer, A. M., Rosen, G. D., & Fitch, R. H. (2002). Sex differences in rapid auditory processing deficits in ectopic BXSB/MpJ mice. *NeuroReport, 13,* 2277–2280.

Peiffer, A. M., Rosen, G. D., & Fitch, R. H. (2004). Sex differences in auditory processing deficits in microgyric rats. *Developmental Brain Research, 148*(1), 53–57.

Persky, R. W., Liu, F., Xu, Y., Weston, G., Levy, S., Roselli, C. E., & McCullough, L. D. (2013). Neonatal testosterone exposure protects adult male rats from stroke. *Neuroendocrinology, 97,* 271–282.

Polyak, A., Rosenfeld, J. A., & Girirajan, S. (2015). An assessment of sex bias in neurodevelopmental disorders. *Genome Medicine, 7,* 94.

Raz, S., Lauterbach, M. D., Hopkins, T. L., Glogowski, B. K., Porter, C. L., Riggs, W.W., & Sander, C. J. (1995). A female advantage in cognitive recovery from early cerebral insult. *Developmental Psychology, 31*(6), 958–966.

Renolleau, S., Fau, S., & Charriaut-Marlangue, C. (2008). Gender-related differences in apoptotic pathways after neonatal cerebral ischemia. *Neuroscientist, 14,* 46–52.

Renolleau, S., Fau, S., Goyenvalle, C., Joly, L-M., Chauvier, D., Jacotot, E.,…Charriaut-Marlangue, C. (2007). Specific caspase inhibitor Q-VD-OPh prevents neonatal stroke in P7 rat: A role for gender. *Journal of Neurochemistry, 100*(4), 1062–1071.

Rucklidge, J. J. (2010). Gender differences in attention-deficit/hyperactivity disorder. *Psychiatric Clinics of North America, 33,* 357–373.

Skirrow, C., Cross, J. H., Cormack, F., Harkness, W., Vargha-Khadem, F., & Baldeweg, T. (2011). Long-term intellectual outcome after temporal lobe surgery in childhood. *Neurology, 76,* 1330–1337.

Smith, A. L., Alexander, M. L., Rosenkrantz, T., & Fitch, R. H. (2014). Sex differences in behavioral outcome following neonatal hypoxia ischemia: Insights from a clinical meta-analysis and a rodent model of induced hypoxic ischemic brain injury. *Experimental Neurology, 254,* 54–67.

Smith, A. L., Rosenkrantz, T. S., & Fitch, R. H. (2016). Neuropathology and neural reorganization following temperature reduction during induced neonatal hypoxic ischemic brain injury in male and female rats. *Neural Plasticity, 2585230.*

Szalkowski, C. E., Booker, A. B., Truong, D. T., Rosen, G. D., & Fitch, R. H. (2013). Knockdown of the candidate dyslexia susceptibility gene homolog Dyx1c1 in rodents: Effects on rapid and complex auditory processing, visual attention, and cortical and thalamic anatomy. *Developmental Neuroscience,* 35(1), 50–68.

Szalkowski, C. E. Hinman, J., Threlkeld, S. W., Wang, Y., LePack, A., Rosen, G. D,…Fitch, R. H. (2011). Persistent spatial working memory deficits in rats following in utero RNAi of Dyx1c1. *Genes, Brain and Behavior, 10,* 244–252.

Taylor, M. J., Lichtenstein, P., Larsson, H., Anckarsäter, H., Greven, C. U., & Ronald, A. (2016). Is there a female protective effect against attention-deficit/hyperactivity disorder? Evidence from two representative twin samples. *Journal of the American Academy of Child and Adolescent Psychiatry, 55,* 504–512.

Threlkeld, S. W., McClure, M. M., Bai, J., Wang, Y., LoTurco, J. J., Rosen, G. D., & Fitch, R. H. (2007). Developmental disruptions and behavioral impairments in rats following in utero RNAi of Dyx1c1. *Brain Research Bulletin, 71*, 508–514.

Turtzo, L. C., & McCullough, L. D. (2008). Sex differences in stroke. *Cerebrovascular Diseases, 26*, 462–474.

Vinet, É., Pineau, C. A., Clarke, A. E., Fombonne, É., Platt, R. W., & Bernatsky, S. (2014). Neurodevelopmental disorders in children born to mothers with systemic lupus erythematosus. *Lupus, 23*, 1099–1104.

Waddell, J., Hanscom, M., Shalon Edwards, N., McKenna, M. C., & McCarthy, M. M. (2016). Sex differences in cell genesis, hippocampal volume and behavioral outcomes in a rat model of neonatal HI. *Experimental Neurology, 275*(Pt 2), 285–295.

Wen, W., & Wen, S. W. (2014). Expanding upon the "extreme male brain" theory of autism as a common link between other major risk factors: A hypothesis. *Medical Hypotheses, 82*, 615–618.

Yang, S. H., Perez, E., Cutright, J., Liu, R., He, Z., Day, A. L., & Simpkins, J. W. (2002). Testosterone increases neurotoxicity of glutamate in vitro and ischemia-reperfusion injury in an animal model. *Journal of Applied Physiology, 92*, 195–201.

Zhu, C., Xu, F., Wang, X., Shibata, M., Uchiyama, Y., Blomgren, K., & Hagberg, H. (2006). Different apoptotic mechanisms are activated in male and female brains after neonatal hypoxia-ischemia. *Journal of Neurochemistry, 96*, 1016–1027.

Zup, S. L., Edwards, N. S., & McCarthy, M. M. (2014). Sex- and age-dependent effects of androgens on glutamate-induced cell death and intracellular calcium regulation in the developing hippocampus. *Neuroscience, 281*, 77–87.

INTEGRATIVE SUMMARY 2

Sex Differences and the Geschwind-Galaburda Hypothesis

Laurie Cutting

SUMMARY

The Geschwind-Galaburda Hypothesis (GGH; Geschwind & Galaburda, 1985a, b, c) proposes that the relationship among genetic factors, prenatal hormonal environment, sex, and brain asymmetries are related to occurrence of dyslexia and, more broadly, to the overrepresentation of males in neurodevelopmental disorders (NDs). The four chapters in this section focus on the central role of sex in the GGH and tackle neurobiological origins of sex difference as related to dyslexia in diverse ways. Overall, examination of the GGH across different neuroimaging modalities, across sex chromosome aneuploidies, and in animal models suggests that the general picture of how sex is related to the dyslexia diagnosis and its neurobiological correlates is very unclear. Several future directions are suggested to address the current limitations of the field, including reporting neuroimaging data broken out by sex, reporting pubertal status of participants, and examining the role of sex and responsiveness to intervention.

INTRODUCTION

Ramus, Altarelli, Jednoróg, Zhao, and di Covella (Chapter 7) examine sex differences in dyslexia across several datasets and different imaging modalities. Results from their studies highlight several issues. First, although some of their findings converge with the GGH, others do not. Ramus and colleagues discuss that some of these differences, however, may stem from methodological differences—the original studies by Geschwind and Galaburda (1985a, b, c) were done on postmortem brains, but moving from postmortem to neuroimaging in vivo is challenging, and the linkages are not one to one. Second, their studies indicate a need to consider what types of measures we are gathering and what that might tell us about when disruption occurs in development. For example, gyri and sulci formation are largely prenatal events (White, Shu, Schmidt, Kao, & Sapior, 2010), suggesting that at least some of the disruption in dyslexia occurs very early in development, at the fetal stage. Third, Ramus and colleagues' findings point toward the need to not simply study the left hemisphere—or even the left and right hemispheres together—but more important, the relationship between the hemispheres, or indices of symmetry, as the critical piece of the puzzle. Finally, Ramus et al. raise the idea that females could have the same biological disruption as males but not be as greatly affected by it. (This idea ties into similar issues raised in Fitch, Bodge, Szalkowski, and Smith, Chapter 10.)

Similarly, Luders (Chapter 8) raises some fundamental methodological questions and highlights the fact that some foundational research may need to be conducted before we can fully understand sex differences as related to dyslexia. Using very careful methodological approaches, she illuminates the shortcomings of some prior conclusions about male–female differences. Her findings reveal that females appear to fundamentally have more gray matter than males, but the two are equal in terms of white matter. Luders points to the intriguing idea that the areas where females have more gray matter overlap with areas known to be estrogen receptor maps, which is consistent with the GGH. The fact that other brain regions not related to estrogen receptors were different between males and females, however, suggests a more complex picture, perhaps one that is revealed by examining both hormones and sex chromosomes together. To this end, Hong's chapter addresses the complexity of each of these factors and how they may influence brain development.

Hong's research (Chapter 9) addresses the GGH by examining sex chromosome aneuploidies, or genetic disorders in which individuals have an abnormal number of sex chromosomes. He illustrates the potential for understanding more about the influence of sex chromosomes on NDs (including dyslexia) through Turner syndrome (TS) and Klinefelter syndrome (KS), which represent the loss (TS) or gain (KS) of genetic material from the X chromosome. These two syndromes result in specific cognitive profiles, with learning disabilities in math for TS and learning disabilities in language-related domains in KS. Hong argues that because the findings of these cognitive profiles are reported in prepubertal individuals, perhaps the differences are not due to hormonal influences during puberty, although prenatal hormonal influences cannot be discounted. He concludes with an important point—although learning disabilities (including dyslexia) could be directly mediated through sex-specific pathways, he notes that the research "suggests that mediation mechanisms alone would not adequately account for observations in LDs [learning disabilities], giving rise to the possibility that sex-specific factors play a moderating role in the relationship between pathophysiological mechanisms of LDs and observed phenotype" (Hong, p. 105). Therefore, although there is evidence that sex is influential in NDs, the mechanisms are not clear at this point.

Although the prior three chapters focus on human imaging and cognitive phenotype, Fitch and colleagues (Chapter 10) shift the focus of sex differences as related to the GGH to animal models (also discussed in Chapter 1). They present intriguing findings that converge with the Luders and Hong chapters in that they show important sex-related differences in mouse models of dyslexia. More important, they concretely highlight the idea that a combination of factors may lead to NDs (dyslexia being one).

The findings Fitch and colleagues review suggest that male–female differences in the prevalence of dyslexia may stem from interactions between genetic and hormonal differences, and females may have the ability to tolerate more disruptions than males. They also raise the intriguing idea that females who do present with severe dyslexia may have a greater load of dyslexia risk factors with only "serious accumulations of risk (injury, mutation, stress, toxins)" leading to a diagnosis of dyslexia in females (Fitch et al., this volume, p. 115). In other words, females may be more resistant to the negative effects of causal factors, so by the time they have enough symptoms to merit the diagnosis of dyslexia, they may have many risk factors simultaneously at play, which is more than what males need for a diagnosis.

Implications

The four chapters have several implications with regard to understanding the impact of sex on dyslexia. The overall picture of how sex is related to the dyslexia diagnosis and its neurobiological correlates is very unclear. This is partly because sex in general has often been ignored in scientific studies (see Cahill, 2006; Clayton & Collins, 2014), but there are also methodological reasons. Each chapter highlights methodological issues and approaches that need to be grappled with in dyslexia research. Ramus et al. (Chapter 7) and Luders (Chapter 8) illustrate the need to more carefully match groups in neuroimaging studies. This has been a significant challenge in the past due to the cost of neuroimaging research. The increasing numbers of pooled datasets that are open to researchers to analyze, however, should help tackle these issues. Hong (Chapter 9) highlights the need for examination of multiple potential influences on brain development, including sex chromosomes and hormonal relationships. Cross-species data interpretation—how to better integrate animal models of NDs with human findings—is another methodological issue. Utilizing neuroimaging techniques in mouse models of NDs would help with this issue. For example, studies have shown that the known disruptions in human brains with Neurofibromatosis Type 1 (NF-1) can also be found in mouse models of NF-1 (e.g., Petrella et al., 2016; Shilyansky et al., 2010). Understanding the relationships between in vivo and ex-vivo findings is yet another methodological issue—animal imaging may bring us closer to understanding because animals can participate in both types of imaging.

Although not explicitly discussed in any of the chapters, a larger methodological issue transcends all of the aforementioned issues—how dyslexia is defined. Ramus et al. and Fitch et al. assume that dyslexia is more common in males than females, which many would agree is the case. Hong, however, points out that not all research agrees on this issue. The Connecticut Longitudinal Study used a random longitudinal sample of 381 children and reported that the ratio of males to females with dyslexia was roughly equal (Shaywitz, Shaywitz, Fletcher, & Escobar, 1990). Nevertheless, other research provided a more nuanced picture. Although a 1:1 ratio exists when dyslexia is defined by a less stringent criteria on reading scores (e.g., 15th–30th percentile) (Quinn & Wagner, 2015; Rutter et al., 2004), studies also show that the male-to-female ratio increases when as you move to greater severity, with a significantly higher male-to-female ratio if dyslexia's reading cutoff criteria score becomes more stringent (e.g., defined by less than or equal to 5th percentile cutoff) (Quinn & Wagner, 2015; Rutter et al., 2004). Trying to pinpoint where dyslexia should be defined on the normal curve is beyond the scope of this chapter, but these findings of increasing male prevalence with greater severity line up nicely with Fitch et al.'s idea about female vulnerability/male vulnerability thresholds. That is, using the higher threshold for dyslexia may be capturing the subthreshold females who have the same biological vulnerability as males, but do not show as severe a phenotype due to protective factors (e.g., when dyslexia is defined as less than or equal to 25th percentile, it may be capturing the subthreshold females with dyslexia that score between 25th and 15th percentiles). Specifically examining whether females at risk for reading difficulties respond to educational interventions either very quickly or extraordinarily slowly is one future research direction that could be of interest. Fitch et al.'s discussion suggests that the subthreshold females would present with a milder phenotype and, therefore, may respond faster, whereas those with a more severe phenotype (e.g., those defined by a 5th percentile cutoff) may have greater neurobiological vulnerability than males, and, therefore, one could hypothesize, would respond more slowly. (Interestingly, administrators from the schools who attended this meeting were queried about this idea and indicated that this was exactly the case as they observe it.)

FUTURE DIRECTIONS

Several areas for future directions could enhance the current state of the science. First, imaging studies and dyslexia studies in general should report data broken out by sex. Sex is rarely addressed in studies of dyslexia, but it is generally controlled for when it is addressed, and the findings are not reported separately for males and females. Reporting findings by sex will allow us to begin to disentangle whether and what types of differences exist between males and females with regard to dyslexia; it would be especially helpful for facilitating meta-analyses focusing on sex differences in dyslexia. Reporting males and females separately is a simple thing to do, and it could be reported as supplementary data in journal articles as not to burden the central components (and eat up page limitations) of researchers' studies. Second, neuroimaging studies should report pubertal status of their participants when possible. This is more complicated because many researchers do not collect these data, but perhaps this could be a consideration for new studies. Third, intervention studies should specifically examine the role of sex and response to intervention (RTI). Although groups are often balanced in terms of sex distribution, it is notable that sex ratios are often not reported in Tier III or IV findings (i.e., the most severe poor readers) in RTI studies. This information could be helpful in understanding whether females represent a different phenotype than males with dyslexia or have a less severe form of dyslexia and accounting for the seemingly greater prevalence of dyslexia in males when dyslexia is defined by more stringent thresholds. Finally, the chapters did not address the issue of postnatal environment and the role it could play in either tempering the potential for developing dyslexia (e.g., by excellent systematic instruction early in development) or exacerbating it (e.g., through impoverished oral language input early in development) when examining the idea of cumulative risk. Future studies will likely need to examine prenatal and postnatal environmental factors as well as a variety of neurobiological and genetic factors to fully understand the role of sex in dyslexia.

REFERENCES

Cahill, L. (2006). Why sex matters for neuroscience. *Nature Neuroscience Reviews, 7,* 477–484.

Clayton, J., & Collins, F. (2014). Policy: NIH to balance sex in cell and animal studies. *Nature, 506,* 282–283. doi:10.1038/509282a

Geschwind, N., & Galaburda, A. M. (1985a). Cerebral lateralization. Biological mechanisms, associations, and pathology: I. A hypothesis and a program for research. *Archives of Neurology, 42*(5), 428–459.

Geschwind, N., & Galaburda, A. M. (1985b). Cerebral lateralization. Biological mechanisms, associations, and pathology: II. A hypothesis and a program for research. *Archives of Neurology, 42*(6), 521–552.

Geschwind, N., & Galaburda, A. M. (1985c). Cerebral lateralization. Biological mechanisms, associations, and pathology: III. A hypothesis and a program for research. *Archives of Neurology, 42*(7), 634–654.

Petrella, L. I., Cai, Y., Sereno, J. V., Gonçalves, S. I., Silva, A. J., & Castelo-Branco, M. (2016). Brain and behaviour phenotyping of a mouse model of neurofibromatosis type-1: An MRI/DTI study on social cognition *Genes, Brain and Behavior, 15*(7), 1–10.

Quinn, J. M., & Wagner, R. K. (2015). Gender differences in reading impairment and in the identification of impaired readers: Results from a large-scale study of at-risk readers. *Journal of Learning Disabilities, 48,* 433–445.

Rutter, M., Caspi, A., Fergusson, D., Horwood, L. J., Goodman, R., Maughan, B.,…Carroll, J. (2004). Sex differences in developmental reading disability: New findings from 4 epidemiological studies. *The Journal of the American Medical Association, 291*(16), 2007–2012. doi:10.1001/jama.291.16.2007

Shaywitz, S. E., Shaywitz, B. A., Fletcher, J. M., & Escobar, M. D. (1990). Prevalence of reading disability in boys and girls. Results of the Connecticut Longitudinal Study. *The Journal of the American Medical Association, 264*(8), 998–1002.

Shilyansky, C., Karlsgodt, K. H., Cummings, D. M., Sidiropoulou, K., Hardt, M., James, A. S.,…Silva, A. J. (2010). Neurofibromin regulates corticostriatal inhibitory networks during working memory performance. *Proceedings of the National Academy of Sciences United States of America,* 107, 13141–13146. doi:10.1073/pnas.1004829107.

White, T., Shu, S., Schmidt, M., Kao, C. Y., & Sapior, G. (2010). The development of gyrification in childhood and adolescence. *Brain and Cognition, 72*(1), 36–45.

SECTION IV

Laterality

CHAPTER 11

From One to Many

Why the Mechanics of Complex Human Traits Are Difficult to Decipher

Elena L. Grigorenko

SUMMARY

One approach to understanding in greater depth the phenomenon of any complex human trait, such as laterality, is to consider the processes and eventualities that underlie its manifestation. Specifically, genetic mosaicism, epigenetics, and the concept of the endophenotype all exemplify the efforts of scientists to identify relevant causal mechanisms. They also emphasize that human complexity is allowed by the variation possible within human systems—what was perceived as one is, in fact, many. Genetic mosaicism is the occurrence of multiple actual genotypes/genomes that operate within the different cells of a single human individual. These are the result of different types of genetic events and account for a portion of individual genetic variation. Epigenetics, the set of molecular processes that influence the expression of the genome, is affected by many environmental factors, both pre- and postnatally and throughout the life span. Multiple epigenomes are created by turning genes on and off in response to environmental cues, leading to greater variation in the ability to adapt to new life conditions. Finally, endophenotypes are the result of parsing phenotypes—that is, dissecting them, often into neurological components—to link the more general phenotypes to their genetic origins. Endophenotypes represent the many components that might contribute to a single clinical phenotype. Thus, each of these concepts explains and exemplifies how complexity may be broken into many parts to ultimately conceive a deeper and fuller vision of a singular phenomenon, such as the occurrence of laterality.

INTRODUCTION

> Our thesis is that although genes contribute importantly, many influences that lie outside of the gene pool of the fetus can alter lateralization patterns. The most powerful factors are variations in the chemical environment in fetal life and to a lesser extent in infancy and early childhood.

The writing of this chapter was supported by Grants R01 DC007665 and R21 HD070594 from the National Institutes of Health, and grant 14.Z50.31.0027 from the Government of the Russian Federation. Grantees undertaking such projects are encouraged to freely express their professional judgment. Therefore, this chapter does not necessarily reflect the position or policies of the abovementioned funding agencies, and no official endorsement should be inferred. I am grateful to Mei Tan for her editorial assistance.

> The factors that modify cerebral dominance also influence the development of many other systems, e.g., the organs involved in the immune response. (Geschwind & Galaburda, 1985, p. 431).

It is well known that humans exhibit structural (e.g., organ based) and functional (e.g., limb and brain based) lateralization in their overall symmetrical bodies. Thus, we are anatomically and phenotypically symmetrical and asymmetrical, and the study of (a)typical development has been focused on understanding (a)typical (a)symmetries and their cross-references in a 2 x 2 table since its inception in ancient times (Sadler, 1997). This has been central for differentiating healthy and disordered development (Brandler & Paracchini, 2014). The strong association between language lateralization in the left hemisphere and right-handedness (Corballis, 2014) is one typical asymmetry, although this association is far from absolute (i.e., it is 96% *vs.* 70% in right and left handers, respectively) (Knecht et al., 2000), and its nature and consequences are hotly disputed (Bishop, 2013).

Laterality is thought to be genetically controlled in all of its manifestations (Brandler & Paracchini, 2014); yet, its mechanisms have been difficult to decipher because of specific mitigating factors. Although lateralization has been typically explained by single-gene theories, the latest data indicate that its mechanisms are more complex (Armour, Davison, & McManus, 2014). Moreover, lateralization is a matter of degree, not category, even though it might appear to be a simple categorical trait; it is a complex phenotype that can be decomposed into or projected onto a number of other phenotypes.

Any doubt about the deterministic role of the gene pool regarding complex human traits has now grown into a deep appreciation of the dynamic pathway that is defined as the initial genome emerges from the unification of a mother's egg and a father's sperm on its way to become the many (epi)genomes in control of a great number of complex human traits. This chapter briefly describes three types of this one-to-many transformations that may occur as the initial genome instantiates a person through the life span and the manifestations of an individual's various complex phenotypes. These transformations reflect the growing appreciation of the nonlinearity of human development and, particularly, the manifestation of complex human traits. Rather than offering a comprehensive literature review of genetic mechanisms of lateralization or other complex traits, my intention is to highlight selected hot spots, indicating where the field will work in the next few years toward elucidating the specific (epi)genetic mechanics of human complex traits.

FROM ONE GENOME TO MORE GENOMES

Genetic mosaicism refers to the presence of two or more genetically distinct cell lines in a single individual that originated from the same zygote. Three types of mosaicism are 1) parental (germline or gonadal) mosaicism, genetic variation in germline cells; 2) somatic mosaicism, variation in somatic cells thought to appear early in development because this period is characterized not only by heightened mutagenic activity but also lowered replication, error checking, and repair activities; and 3) gonosomal mosaicism, in which genetic variation is present in both somatic and germline cells (Campbell, Shaw, Stankiewicz, & Lupski, 2015). Sporadic de novo mosaicism occurs in only one particular individual.

As somatic mutations arise primarily during cell division, cells that naturally divide more frequently due to the specificity (e.g., epithelial cells, hematopoietic cells, male germ cells) of their corresponding tissues are more susceptible than cells in tissues that rarely divide (Campbell et al., 2015). Given that an adult human has approximately 10^{14} cells that resulted from approximately 10^{16} mitoses (Iourov, Vorsanova, & Yurov, 2010), each of us likely has numerous cells that harbor causative factors for many or even all human genetic maladies; yet, only some of these mutations have been shown to be impactful. Indeed, some thought that somatic mutations were limited to the lifetime of the individual (i.e., not transmitted to the next generation) (De, 2011). It has now been demonstrated, however, that somatic mutations originating during the early postzygotic stage of embryonic development might result in mosaicism with subsequent transmission of the mutation through the germ cell (Campbell et al., 2015). It also appears that somatic mutations occur at an increased frequency with age (Holstege et al., 2014). Somatic mosaicism results from genetic/genomic modifications of different kind and size (Campbell et al., 2015): aneuploidy (e.g., trisomy 21), large-scale structural abnormalities (e.g., isochromosome 12p), copy number variants (CNVs), single nucleotide variants (SNVs), small insertions and deletions, repeat expansion and contraction, and the insertions of autonomous mobile elements. More important, accumulating evidence suggests that mosaicism is more widespread than earlier thought (Donkervoort et al., 2015).

The potential mechanistic value of somatic mosaicism for human development in general and the manifestation of neurodevelopmental disorders (NDs) in particular are still difficult to precisely appraise, but a number of observations in the literature underscore their relevance as the source of individual differences and the magnitude of their effect. For example, reports of monozygotic twins (MZ) discordant for various genomic lesions such as aneuploidy of chromosomes X and Y (Razzaghian et al., 2010), noncoding fragile X mental retardation 1 (FMR1) repeat length (Helderman-van Den Enden et al., 1999), and different copy number profiles (Abdellaoui et al., 2015), especially while aging (Bruder et al., 2008), have powerful developmental and phenotypic implications. Yet, reports suggest that somatic mutational events in coding regions are rare and might not have a major role in the phenotypic expression of MZ discordant twin pairs (Lyu et al., 2016; Zhang et al., 2016).

The very presence of tissue-specific mosaicism introduces a problem in understanding the etiology of NDs by focusing on a single cell type, which is typically either blood or saliva. In fact, causal mutations might be present in cells derived from other lineages that were not analyzed simply due to their unavailability (Donkervoort et al., 2015). This is particularly relevant when the etiology of a disorder involves brain cells, as all NDs are thought to do. Although limited, the literature contains a few relevant illustrations. In general, somatic mosaicism appears to be an important mechanism in the manifestation of NDs (Abyzov et al., 2012). For example, adult cortical neurons appear to be somatically mosaic for insertions of mobile elements—retrotransposons (Erwin, Marchetto, & Gage, 2014)—with the estimates for the scale of such insertions being highly variable (i.e., from 0.6 [Evrony et al., 2012] to 800 [Coufal et al., 2009] insertions per neuron). Cai and colleagues (2014) used single cell analysis techniques to study samples of brain tissue from individuals with a number of brain malformations and revealed the presence of somatic CNVs, which were shared by multiple neurons and formed changes likely to result in brain pathology, even when the entire brain was not affected. Jamuar and colleagues (2014) looked at high coverage sequencing of candidate genes associated with particular types of cerebral cortical malformations and detected somatic mutations in 8 of the 27 (30%) patients.

FROM THE GENOME TO THE EPIGENOME

Epigenetic processes (also referred to as the epigenome) are a family of processes that regulate gene transcription and maintain genome integrity. They can be classified into a number of categories by their target (e.g., DNA, histones, chromatin), mechanism (e.g., methylation, acetylation, phosphorylation, regulation by noncoding RNAs—nc RNAs), timing (e.g., pretranscriptionally—through defining regions of DNA that need to be transcribed, postranscriptionally—through ncRNAs), and origin (e.g., maintained as in the action of DNA methyltransferase 1 [DNMT1], de novo, as in the action of DNMT3a and 3b). Epigenetic processes vary in their longevity of impact, resulting in both transient (e.g., eroding, erasable) and permanent (e.g., lasting for lifetime, heritable) changes. Although these processes are often studied in isolation, they concurrently and sequentially comodulate gene transcription and translation. The main purpose of epigenetic processes is to confer cell-specific gene expression identity that is established during embryonic development (Lister et al., 2009). Yet, they fulfill many other goals, collectively regulating the function of the genome as it exists in and responds to specific environments, inserting themselves between the gene pool and emerging phenotypes.

After years of in vitro and model organism research on the role of epigenetic processes, a focus on their manifestation and control of complex human behaviors has generated several observations. First, the epigenomes of MZ twins are similar during the early years of life but begin to differentiate in childhood, and the rate and magnitude of differentiation increases with degree of discordance in lifestyle and with age (e.g., Lévesque et al., 2014). Second, numerous epigenetic processes involve various nutrients, making diet, nutrition, and substance intake highly influential on epigenetic status (Haggarty, 2013). The methylation pathway depends on vitamins B and methionine, whereas acetylation depends on carbohydrates, fats, and some amino acids (Haggarty, 2015). The reported association with epigenetic profiles is not surprising (McGuinness et al., 2012), given the socioeconomic stratification of dietary patterns. Third, the sensitivity of the human epigenome to environmental influences appears to be heterogeneous across the life span, with activity especially high in the gametes before fertilization and in the early embryo (Lane, Robker, & Robertson, 2014), but it is also substantial early in life (Naumova, Dozier et al., under review). Little is known about the rate, number, and amount of normative epigenetic changes in human development, however. Fourth, the connections among early life stressors (ELS), epigenetic processes, and subsequent physical and mental health are a dominant focus in epigenetics today.

ELS refer to a wide range of exogenous phenomena that can occur during the fetal stage of life (e.g., maternal malnutrition, elevated glucocorticoid exposure resulting from maternal stress) through childhood and adolescence (e.g., child abuse and neglect, poverty, orphanhood, exposure to violence). Since the late 1990s, evidence has accumulated indicating the major role of epigenetic mechanisms in processes connecting ELS to later physical (e.g., Danese & Tan, 2014) and mental (Naumova, Hein et al., 2016) health. Human research is predominantly (if not exclusively) correlational, associating ELS with epigenetic processes in the brain and peripheral tissues; however, when considered together with the animal literature, there is evidence of an epigenetic component in the biological embedding of ELS that contributes to or even shapes risk and resilience factors for the subsequent development and manifestation of disordered phenotypes. ELS-associated changes in the epigenome are viewed as causally linked to the increase in risk for cardiovascular and metabolic disorders (Danese & McEwen, 2012), proinflammatory immune profiles in adulthood (Danese, Pariante, Caspi, Taylor, & Poulton, 2007), heightened body mass index (Danese & Tan, 2014), and depression (McGowan et al., 2009), among other challenges to human physical and mental health (Provençal & Binder, 2015). Fifth, the list of physical (e.g., Chang & Neu, 2015) and psychiatric disorders (e.g., Nieto, Patriquin, Nielsen, & Kosten, 2016) reportedly related to alterations to the epigenome grows daily, regardless of the etiology.

The rapidly growing literature on the epigenome (or epigenomes, as every individual possesses tissue-specific epigenomes) convincingly demonstrates its role in a series of feedback loops connecting the human genome to the environments in which it exists. Yet, the field is only beginning to understand the verticality of the epigenome (or system of epigenomes possessed by a single individual) and the implication of its (mal)functioning for NDs.

FROM ONE PHENOTYPE TO MANY ENDOPHENOTYPES

The notion that NDs have a genetic etiology is longstanding and widely accepted. This notion has grown more complex (and, probably, more realistic), however, in the 100–150 years since the differentiation of clinical practice and scientific inquiry in developmental disabilities. In fact, this notion suggests that it is not the disorder itself (i.e., a single phenotype, however defined, with the example of laterality, initially thought of as a simple category that turned out to be a complex multilevel human trait) that is controlled genetically, but rather a set of subphenotypes (endophenotypes) that are also controlled genetically and are perhaps closer to the genome than the target phenotype itself. Accordingly, a disorder arises as a derivative of a set of endophenotypes, some of which are atypical, that can manifest as different traits (e.g., electrophysiological, biochemical, endocrinological, anatomical); crucially, however, they are 1) directly related to the target phenotype (i.e., a neurodevelopmental disorder) and 2) heritable in the general population.

The concept of an endophenotype was initially proposed in the context of research on schizophrenia (Gottesman & Shields, 1972) as a means of reducing phenotypic complexity, homogenizing phenotypic variability, and magnifying genetic effects (Bearden & Freimer, 2006), with the belief that the dissection of single complex phenotypes of developmental disabilities into multiple endophenotypes would aid the discovery of the genetic mechanisms underlying the manifestation of these disabilities. Such discovery is assumed to be possible via approaches that are either forward (i.e., from genes to endophenotypes) or reversed (i.e., from endophenotypes to the genome) (Meyer-Lindenberg & Weinberger, 2006). Forward approaches are typically classified by how the targeted genes are identified, either through a theory- or literature-search-driven selection of reasonable candidate genes or through follow-up investigations of specific genes and gene pathways identified through whole-genome association studies (GWAS). Reverse approaches start with the endophenotypes and either subject them to GWAS or include a whole-brain versus whole-genome hypotheses-free interrogation. The most desired and investigated sets of endophenotypes for studies of neuropsychiatric disorders are those pertaining to the brain's structure and activity as elicited either in samples of individuals with target disorders or while studying brain activity during the performance of specific tasks thought to be informative in differentiating individuals who are healthy from individuals affected with disease. In fact, a new bourgeoning area of research—imaging genetics—focuses specifically on understanding the three-level connections among the genome, the brain, and human disorders.

As the field's knowledge of the genome accumulates, there are increasing attempts to investigate associations between structural variation in genes that are considered reasonable candidates for particular endophenotypes and targeted phenotypes. The reasonableness of candidates is generally determined by the field's total knowledge about their function; however, apparent reasonableness has not necessarily translated into statistically significant findings because the field is replete with nonreplications (Chabris et al., 2012).

The greatest success, although not without complexities (Trivedi et al., 2006), is registered in Alzheimer's disease, in which the first study thought to have triggered the development of the field of imaging genetics (Bookheimer et al., 2000) was followed by a set of studies that confirmed and further developed the original findings (e.g., Wishart et al., 2006). Specifically, certain patterns of brain activation are registered during memory tasks for participants differing by genotype at two specific sites within the *APOE* gene—a candidate gene that has been reliably and consistently associated with cognitive functions, explaining ~2% of the variance (Wisdom, Callahan, & Hawkins, 2011). The E4 genotype, which is a version of the *APOE* gene, has been implicated in Alzheimer's disease. E4 carriers demonstrated greater brain activity in the hippocampus when compared with noncarriers, suggesting that such hyperactivity could be a compensatory mechanism assisting task performance that is manifested prior to the onset of the disease. Utilizing genetic variants identified through GWAS is another type of forward approach in imaging genetic studies. One such GWAS-based variant—in the *ZNF804A* gene for schizophrenia (O'Donovan et al., 2008)—has been interrogated within a neuroimaging framework (Esslinger et al., 2009), demonstrating that its genotype differentiated the connectivity between the DLPFC and hippocampal formation, possibly identifying an endophenotype whose particular values may be viewed as risk biomarkers of the disorder (Rasetti et al., 2011). A variation of this approach focuses on a brainwide phenotype rather than singling out a particular endophenotype from the many location- or structure-specific phenotypes that can be generated in neuroimaging studies (e.g., Braskie et al., 2011).

The reversed approach assumes either a focus on a candidate phenotype (measured by neuroimaging) with a GWAS or an adaptation of a hypothesis-free approach in which a comparison of the entire genome with the entire brain in large datasets takes place. The first has been illustrated in a number of GWAS, focusing on a range of brain-based phenotypes. These include both structural (e.g., the volume of the hippocampus [Potkin et al., 2009] and the volume of the caudate [Stein et al., 2011]) and functional phenotypes (e.g., such as activity-based measures of DLPFC during working memory [Papassotiropoulos et al., 2013]), hippocampal activity during episodic memory (Papassotiropoulos et al., 2006), or amygdala reactivity during an emotional faces task (Ousdal et al., 2012). As with all GWAS, the replication and reproducibility of these results is an issue (e.g., Guella et al., 2013). The second approach combines GWAS and brainwide scanning (Linden, 2012).

Clearly, any connection between the genome and NDs is routed through the brain. The range of endophenotypes relevant to the manifestation of any neurodevelopmental disorder is difficult to estimate but must clearly be large. Indeed, no direct correspondence is found among lesions in the genome, lesions in the brain, and lesions in behavior, however these lesions are defined. A count listed 100 small-effect genes collectively explaining ~20% of the genetic variance in schizophrenia (Ripke et al., 2013). There is also evidence for a substantial polygenic component to variation in general cognitive ability in the general population (Davies et al., 2011). Moreover, there is evidence for shared genetic variance between general cognitive ability and schizophrenia (McIntosh et al., 2013). There are brain endophenotypes that are likely to substantiate this shared variance (McIntosh, Deary, & Porteous, 2014).

CONCLUSION

This chapter began with a brief overview of the state of affairs for understanding the genetic mechanisms of lateralization. Lateralization, highlighted in the interpretation of Geschwind and Galaburda (1987) as a key trait manifested early in development and relevant to many downstream developmental events, from sex differences to brain-based disorders, is etiologically complex. My contribution has been to suggest why this might be the case. Multiple epigenomes are created by turning genes on and off in response to environmental cues, leading to greater variation in the ability to adapt to new life conditions. Endophenotypes are the result of parsing phenotypes to link the more general phenotypes to their genetic origins and represent the components that might contribute to a single clinical phenotype. Thus, each of these concepts explains and exemplifies how complexity may be broken into many parts to ultimately conceive a deeper and fuller vision of a singular phenomenon, such as the occurrence of laterality.

REFERENCES

Abdellaoui, A., Ehli, E. A., Hottenga, J. J., Weber, Z., Mbarek, H., Willemsen, G.,… Boomsma, D. I. (2015). CNV concordance in 1,097 MZ twin pairs. *Twin Research and Human Genetics, 18,* 1–12. doi:10.1017/thg.2014.86

Abyzov, A., Mariani, J., Palejev, D., Zhang, Y., Haney, M. S., Tomasini, L.,…Vaccarino, F. M. (2012). Somatic copy-number mosaicism in human skin revealed by induced pluripotent stem cells. *Nature, 492,* 438–442.

Armour, J. A., Davison, A., & McManus, I. C. (2014). Genome-wide association study of handedness excludes simple genetic models. *Heredity, 112,* 221–225. doi:10.1038/hdy.2013.93

Bearden, C. E., & Freimer, N. B. (2006). Endophenotypes for psychiatric disorders: ready for primetime? *Trends in Genetics, 22,* 306–313.

Bishop, D. V. M. (2013). Cerebral asymmetry and language development: Cause, correlate, or consequence? *Science, 340,* 1230531. doi:10.1126/science.1230531

Bookheimer, S. Y., Strojwas, M. H., Cohen, M. S., Saunders, A. M., Pericak-Vance, M. A., Mazziotta, J. C., & Small, G. W. (2000). Patterns of brain activation in people at risk for Alzheimer's disease. *New England Journal of Medicine, 343,* 450–456. doi:10.1056/NEJM200008173430701

Brandler, W. M., & Paracchini, S. (2014). The genetic relationship between handedness and neurodevelopmental disorders. *Trends in Molecular Medicine, 20,* 83–90. doi:10.1016/j.molmed.2013.10.008

Braskie, M. N., Jahanshad, N., Stein, J. L., Barysheva, M., McMahon, K. L., de Zubicaray, G. I.,...Thompson, P. M. (2011). Common Alzheimer's disease risk variant within the CLU gene affects white matter microstructure in young adults. *Journal of Neuroscience, 31,* 6764–6770. doi:10.1523/JNEUROSCI.5794-10.2011

Bruder, C. E. G., Piotrowski, A., Gijsbers, A. A. C. J., Andersson, R., Erickson, S., Diaz de Ståhl, T.,...Dumanski, J. P. (2008). Phenotypically concordant and discordant monozygotic twins display different DNA copy-number-variation profiles. *American Journal of Human Genetics, 82*: 763–771. doi:10.1016/j.ajhg.2007.12.011

Cai, X., Evrony, G. D., Lehmann, H. S., Elhosary, P. C., Mehta, B. K., Poduri, A., & Walsh, C. A. (2014). Single-cell, genome-wide sequencing identifies clonal somatic copy-number variation in the human brain. *Cell reports, 8,* 1280–1289. doi:10.1016/j.celrep.2014.07.043

Campbell, I. M., Shaw, C. A., Stankiewicz, P., & Lupski, J. R. (2015). Somatic mosaicism: Implications for disease and transmission genetics. *Trends in Genetics, 31,* 382–392. doi:10.1016/j.tig.2015.03.013

Chabris, C. F., Hebert, B. M., Benjamin, D. J., Beauchamp, J., Cesarini, D., van der Loos, M.,...Laibson, D. (2012). Most reported genetic associations with general intelligence are probably false positives. *Psychological Science, 23,* 1314–1323.

Chang, L., & Neu, J. (2015). Early factors leading to later obesity: Interactions of the microbiome, epigenome, and nutrition. *Current Problems in Pediatric and Adolescent Health Care, 45,* 134–142. doi:http://dx.doi.org/10.1016/j.cppeds.2015.03.003

Corballis, M. C. (2014). Left brain, right brain: Facts and fantasies. *PLoS Biology, 12,* e1001767. doi:10.1371/journal.pbio.1001767

Coufal, N. G., Garcia-Perez, J. L., Peng, G. E., Yeo, G. W., Mu, Y., Lovci, M. T.,...Gage, F. H. (2009). L1 retrotransposition in human neural progenitor cells. *Nature, 460,* 1127–1131. doi:http://www.nature.com/nature/journal/v460/n7259/suppinfo/nature08248_S1.html

Danese, A., & McEwen, B. S. (2012). Adverse childhood experiences, allostasis, allostatic load, and age-related disease. *Physiology and Behavior, 106,* 29–39. doi:http://dx.doi.org/10.1016/j.physbeh.2011.08.019

Danese, A., Pariante, C. M., Caspi, A., Taylor, A., & Poulton, R. (2007). Childhood maltreatment predicts adult inflammation in a life-course study. *Proceedings of the National Academy of Sciences USA, 104,* 1319–1324. doi:10.1073/pnas.0610362104

Danese, A., & Tan, M. (2014). Childhood maltreatment and obesity: Systematic review and meta-analysis. *Molecular Psychiatry, 19,* 544–554. doi:10.1038/mp.2013.54

Davies, G., Tenesa, A., Payton, A., Yang, J., Harris, S. E., Liewald, D.,...Deary, I. J. (2011). Genome-wide association studies establish that human intelligence is highly heritable and polygenic. *Molecular Psychiatry, 16,* 996–1005. doi:10.1038/mp.2011.85

De, S. (2011). Somatic mosaicism in healthy human tissues. *Trends in Genetics, 27,* 217–223.

Donkervoort, S., Hu, Y., Stojkovic, T., Voermans, N. C., Foley, A. R., Leach, M. E.,...Bönnemann, C. G. (2015). Mosaicism for dominant collagen 6 mutations as a cause for intrafamilial phenotypic variability. *Human Mutation, 36,* 48–56.

Erwin, J. A., Marchetto, M. C., & Gage, F. H. (2014). Mobile DNA elements in the generation of diversity and complexity in the brain. *Nature Reviews Neuroscience, 15,* 497–506. doi:10.1038/nrn3730

Esslinger, C., Walter, H., Kirsch, P., Erk, S., Schnell, K., Arnold, C.,...Meyer-Lindenberg, A. (2009). Neural mechanisms of a genome-wide supported psychosis variant. *Science, 324,* 605. doi:10.1126/science.1167768

Evrony, G. D., Cai, X., Lee, E., Hills, L. B., Elhosary, P. C., Lehmann, H. S.,…Walsh, C. A. (2012). Single-neuron sequencing analysis of L1 retrotransposition and somatic mutation in the human brain. *Cell, 151,* 483–496. doi:http://dx.doi.org/10.1016/j.cell.2012.09.035

Geschwind, N., & Galaburda, A. M. (1985). Cerebral lateralization. Biological mechanisms, associations, and pathology: I. A hypothesis and a program for research. *Archives of Neurology, 42*(5), 428–459.

Geschwind, N., & Galaburda, A. M. (1987). *Cerebral lateralization: biological mechanisms, associations and pathology.* Cambridge, MA: The MIT Press.

Gottesman, I. I., & Shields, J. (1972). *Schizophrenia and genetics: A twin study vantage point.* New York, NY: Academic Press.

Guella, I., Sequeira, A., Rollins, B., Morgan, L., Torri, F., van Erp, T. G. M.,…Vawter, M. P. (2013). Analysis of miR-137 expression and rs1625579 in dorsolateral prefrontal cortex. *Journal of psychiatric research, 47,* 1215–1221. doi:http://dx.doi.org/10.1016/j.jpsychires.2013.05.021

Haggarty, P. (2013). Epigenetic consequences of a changing human diet. *Proceedings of the Nutrition Society, 72,* 363–371. doi:10.1017/S0029665113003376

Haggarty, P. (2015). Genetic and metabolic determinants of human epigenetic variation. *Current Opinion in Clinical Nutrition and Metabolic Care, 18,* 334–338. doi:10.1097/mco.0000000000000194

Helderman-van Den Enden, A. T. J. M., Maaswinkel-Mooij, P. D., Hoogendoorn, E., Willemsen, R., Maat-Kievit, J. A., Losekoot, M., & Oostra, B. A. (1999). Monozygotic twin brothers with the fragile X syndrome: Different CGG repeats and different mental capacities. *Journal of Medical Genetics, 36,* 253–257.

Holstege, H., Pfeiffer, W., Sie, D., Hulsman, M., Nicholas, T. J., Lee, C. C.,…Sistermans, E. A. (2014). Somatic mutations found in the healthy blood compartment of a 115-yr-old woman demonstrate oligoclonal hematopoiesis. *Genome Research, 24,* 733–742. doi:10.1101/gr.162131.113

Iourov, I. Y., Vorsanova, S. G., & Yurov, Y. B. (2010). Somatic genome variations in health and disease. *Current Genomics, 11,* 387–396.

Jamuar, S. S., Lam, A.-T. N., Kircher, M., D'Gama, A. M., Wang, J., Barry, B. J.,…Walsh, C. A. (2014). Somatic mutations in cerebral cortical malformations. *New England Journal of Medicine, 371,* 733–743. doi:10.1056/NEJMoa1314432

Knecht, S., Dräger, B., Deppe, M., Bobe, L., Lohmann, H., Flöel, A.,…Henningsen, H. (2000). Handedness and hemispheric language dominance in healthy humans. *Brain, 123,* 2512–2518. doi:10.1093/brain/123.12.2512

Lane, M., Robker, R. L., & Robertson, S. A. (2014). Parenting from before conception. *Science, 345,* 756–760.

Lévesque, M. L., Casey, K. F., Szyf, M., Ismaylova, E., Ly, V., Verner, M.-P.,…Booij, L. (2014). Genome-wide DNA methylation variability in adolescent monozygotic twins followed since birth. *Epigenetics, 9,* 1410–1422. doi:10.4161/15592294.2014.970060

Linden, D. E. (2012). The challenges and promise of neuroimaging in psychiatry. *Neuron, 73,* 8–22.

Lister, R., Pelizzola, M., Dowen, R. H., Hawkins, R. D., Hon, G., Tonti-Filippini, J.,…Ecker, J. R. (2009). Human DNA methylomes at base resolution show widespread epigenomic differences. *Nature, 462,* 315–322. doi:http://www.nature.com/nature/journal/v462/n7271/suppinfo/nature08514_S1.html

Lyu, N., Guan, L. L., Ma, H., Wang, X. J., Wu, B. M., Shang, F. H.,…Yu, X. (2016). Failure to identify somatic mutations in monozygotic twins discordant for schizophrenia by whole exome sequencing. *Chinese Medical Journal, 129,* 690–695. doi:10.4103/0366-6999.178009

McGowan, P. O., Sasaki, A., D'Alessio, A. C., Dymov, S., Labonte, B., Szyf, M.,…Meaney, M. J. (2009). Epigenetic regulation of the glucocorticoid receptor in human brain associates with childhood abuse. *Nature Neuroscience, 12,* 342–348.

McGuinness, D., McGlynn, L. M., Johnson, P. C. D., MacIntyre, A., Batty, G. D., Burns, H.,…Shiels, P. G. (2012). Socio-economic status is associated with epigenetic differences in the pSoBid cohort. *International Journal of Epidemiology, 41,* 151–160. doi:10.1093/ije/dyr215

McIntosh, A. M., Deary, I. J., & Porteous, D. J. (2014). Two-back makes step forward in brain imaging genomics. *Neuron, 81,* 959–961. doi:http://dx.doi.org/10.1016/j.neuron.2014.02.023

McIntosh, A. M., Gow, A., Luciano, M., Davies, G., Liewald, D. C., Harris, S. E.,...Deary, I. J. (2013). Polygenic risk for schizophrenia is associated with cognitive change between childhood and old age. *Biological Psychiatry, 73,* 938–943. doi:http://dx.doi.org/10.1016/j.biopsych.2013.01.011

Meyer-Lindenberg, A., & Weinberger, D. R. (2006). Intermediate phenotypes and genetic mechanisms of psychiatric disorders. *Nature Review Neuroscience, 7,* 818–827.

Naumova, O. Y., Dozier, D., Dobrynin, P. V., Grigorev, K., Wallin, A., Jeltova, I.,...Grigorenko, E. L. (under review). *Developmental dynamics of the epigenome: A longitudinal study of three toddlers.*

Naumova, O. Y., Hein, S., Suderman, M., Barbot, B., Lee, M., Raefski, A.,...Grigorenko, E. L. (2016). Epigenetic patterns modulate the connection between developmental dynamics of parenting and offspring psychosocial adjustment. *Child Development, 87,* 98–110. doi:doi: 10.1111/cdev.12485

Nieto, S. J., Patriquin, M. A., Nielsen, D. A., & Kosten, T. A. (2016). Don't worry, be informed about the epigenetics of anxiety. *Pharmacology Biochemistry and Behavior, 146–147,* 60–72. doi:http://dx.doi.org/10.1016/j.pbb.2016.05.006

O'Donovan, M. C., Craddock, N., Norton, N., Williams, H., Peirce, T., Moskvina, V.,...Owen, M. J. (2008). Identification of loci associated with schizophrenia by genome-wide association and follow-up. *Nature Genetics, 40,* 1053–1055. doi:http://www.nature.com/ng/journal/v40/n9/suppinfo/ng.201_S1.html

Ousdal, O. T., Anand Brown, A., Jensen, J., Nakstad, P. H., Melle, I., Agartz, I.,...Andreassen, O. A. (2012). Associations between variants near a monoaminergic pathways gene (PHOX2B) and amygdala reactivity: A genome-wide functional imaging study. *Twin Research and Human Genetics, 15,* 273–285.

Papassotiropoulos, A., Stefanova, E., Vogler, C., Gschwind, L., Ackermann, S., Spalek, K.,...de Quervain, D. J. F. (2013). A genome-wide survey and functional brain imaging study identify CTNNBL1 as a memory-related gene. *Molecular Psychiatry, 18,* 255–263. doi:10.1038/mp.2011.148

Papassotiropoulos, A., Stephan, D. A., Huentelman, M. J., Hoerndli, F. J., Craig, D. W., Pearson, J. V.,...de Quervain, D. J.-F. (2006). Common Kibra alleles are associated with human memory performance. *Science*(5798), 475–478. doi:10.1126/science.1129837

Potkin, S. G., Guffanti, G., Lakatos, A., Turner, J. A., Kruggel, F., Fallon, J. H.,...Initiative for the Alzheimer's Disease Neuroimaging. (2009). Hippocampal atrophy as a quantitative trait in a genome-wide association study identifying novel susceptibility genes for Alzheimer's Disease. *PLoS One, 4,* e6501. doi:10.1371/journal.pone.0006501

Provençal, N., & Binder, E. B. (2015). The effects of early life stress on the epigenome: From the womb to adulthood and even before. *Experimental Neurology, 268,* 10–20. doi:http://dx.doi.org/10.1016/j.expneurol.2014.09.001

Rasetti, R., Sambataro, F., Chen, Q., Callicott, J. H., Mattay, V. S., & Weinberger, D. R. (2011). Altered cortical network dynamics: A potential intermediate phenotype for schizophrenia and association with znf804a. *Archives of General Psychiatry, 68,* 1207–1217. doi:10.1001/archgenpsychiatry.2011.103

Razzaghian, H. R., Shahi, M. H., Forsberg, L. A., de Ståhl, T. D., Absher, D., Dahl, N.,...Dumanski, J. P. (2010). Somatic mosaicism for chromosome X and Y aneuploidies in monozygotic twins heterozygous for sickle cell disease mutation. *American Journal of Medical Genetics Part A, 152A,* 2595–2598. doi:10.1002/ajmg.a.33604

Ripke, S., O'Dushlaine, C., Chambert, K., Moran, J. L., Kähler, A. K., Akterin, S.,...Sullivan, P. F. (2013). Genome-wide association analysis identifies 14 new risk loci for schizophrenia. *Nature Genetics, 45,* 1150–1159. doi:10.1038/ng.2742

Sadler, N. (1997). A sinister way of life: A search for left-handed material culture. In S. M. Pearce (Ed.), *Experiencing material culture in the Western world* (pp. 140–153). London, UK: Leicester University Press.

Stein, J. L., Hibar, D. P., Madsen, S. K., Khamis, M., McMahon, K. L., de Zubicaray, G. I.,... Alzheimer's Disease Neuroimaging Initiative. (2011). Discovery and replication of dopamine-related gene effects on caudate volume in young and elderly populations (N = 1198) using genome-wide search. *Molecular Psychiatry, 16,* 927–937. doi:10.1038/mp.2011.32

Trivedi, M. A., Schmitz, T. W., Ries, M. L., Torgerson, B. M., Sager, M. A., Hermann, B. P.,...Johnson, S. C. (2006). Reduced hippocampal activation during episodic encoding in middle-aged individuals at genetic risk of Alzheimer's disease: A cross-sectional study. *BMC Medicine, 4,* 1–14. doi:10.1186/1741-7015-4-1

Wisdom, N. M., Callahan, J. L., & Hawkins, K. A. (2011). The effects of apolipoprotein E on non-impaired cognitive functioning: A meta-analysis. *Neurobiology of Aging, 32,* 63–74. doi:http://dx.doi.org/10.1016/j.neurobiolaging.2009.02.003

Wishart, H. A., Saykin, A. J., Rabin, L. A., Santulli, R. B., Flashman, L. A., Guerin, S. J.,... McAllister, T. W. (2006). Increased brain activation during working memory in cognitively intact adults with the APOE ε4 allele. *American Journal of Psychiatry, 163,* 1603–1610. doi:doi:10.1176/ajp.2006.163.9.1603

Zhang, R., Thiele, H., Bartmann, P., Hilger, A. C., Berg, C., Herberg, U.,...Reutter, H. (2016). Whole-exome sequencing in nine monozygotic discordant twins. *Twin Research and Human Genetics, 19,* 60–65. doi:10.1017/thg.2015.93

CHAPTER 12

The Ontogenesis of Handedness and Language Lateralization

Links to Developmental Disorders

Sebastian Ocklenburg and Jutta Peterburs

SUMMARY

This chapter focuses on recent understanding of the genetic factors that underlie hemispheric dominance, specifically handedness and left-hemisphere language dominance. No single gene shows a significant association for both traits, and the neurophysiological bases for these traits differ; the traits show not only some shared but also substantial independent genetic influences. This is important because the high prevalence of atypical asymmetries in certain mental and developmental disorders suggests that the genetic factors contributing to hemispheric asymmetries might also influence the development of these disorders. Grounded in a symptom-based approach to the genetic contributions to schizophrenia, we propose applying this approach to understand the relation of atypical lateralization and developmental disorders such as dyslexia. The Geschwind-Galaburda Hypothesis (GGH; Geschwind & Galaburda, 1985a, b, c) assumes that differences in intrauterine testosterone levels and sensitivity to testosterone affect brain structure, functional lateralization, immune parameters, and learning disorders (LDs). The human genome had not yet been sequenced and many molecular techniques available today were unavailable when the GGH was first presented in the mid-1980s. Many of its theoretical constructs have not withstood the test of time, but the core assumption concerning lateralization—it is determined to a large extent by nongenetic factors and to a small extent by genetic factors—is largely supported by research. Given the current theory limbo of lateralization research, the core assumptions of the GGH provide a great foundation to develop a multifactorial model. Careful integration of cutting-edge neurogenetic evidence into the core constructs of the GGH could result in a GGH 2.0 that will hopefully inspire as many researchers over the next 30 years as the original has done since the mid-1980s.

INTRODUCTION

The GGH (Geschwind & Galaburda, 1985a, b, c) assumes that functional hemispheric asymmetries are determined by structural hemispheric asymmetries. Their ontogenesis is driven by genetic factors and (to a larger extent) nongenetic factors that modify the direction and extent of structural differences both before and after birth. Although the nongenetic factors affecting lateralization have been laid out in detail elsewhere (Geschwind & Galaburda, 1985a, b, c), this chapter focuses on recent advances in understanding the genetic underpinnings of lateralization. It specifically focuses on handedness and dominance of the left hemisphere for speech production and perception, probably the best-known examples for functional hemispheric asymmetries in humans. The relevance of this research for understanding psychiatric and developmental disorders is investigated in the second part of the chapter.

THE GENETICS OF HANDEDNESS AND LANGUAGE LATERALIZATION

The GGH (Geschwind & Galaburda, 1985a, b, c) showed that direct genetic effects account for a smaller proportion of the variance in handedness data (about 25%) than nongenetic effects (about 75%) (Medland et al., 2009). Nevertheless, as direct genetic effects still account for a considerable amount of shared variance, researchers have tried to identify genes that are relevant for handedness and other forms of hemispheric asymmetries.

Historically, it has been assumed that handedness and language lateralization are determined by the same single gene (e.g., Annett, 1996) because the two traits are correlated to some extent. Evidence strongly argues against a single gene determining functional lateralization, however. For example, Eriksson et al. (2010) conducted a genome-wide association study (GWAS) in which they did not observe any significant associations with handedness in more than 0.5 million single nucleotide polymorphisms (SNPs). This finding was replicated in an independent GWAS by Armour, Davison, and McManus (2014). Similarly, Kavaklioglu, Ajmal, Hameed, and Francks (2015), who conducted whole-exome sequencing in a Pakistani family with higher incidence of left-handedness and a history of consanguineous marriages leading to a potential genetic bottleneck for asymmetry genes, did not observe any candidate variations or rare mutations significantly associated with handedness. This is also in accordance with a linkage analysis in a Dutch population isolate (Somers et al., 2015). These findings, however, do not imply that genetic influences do not play a role for handedness or other forms of hemispheric asymmetries. About 30–100 genes are involved, and a sample size of more than 250,000 participants would be needed to gain appropriate statistical power to identify these genes and their interactions on the genome-wide level (Armour et al., 2014; McManus, Davison, & Armour, 2013). Thus, we are looking for multiple genes with a low amount of explained variance each. Several genes have been identified that might be relevant for handedness (see Table 12.1) and language lateralization (see Table 12.2).

Table 12.1. Genes associated with handedness

Gene name	Location	Study
LMO4	1p22.3	Li et al. (2013)
LRRTM1	2p12	Francks et al. (2007)
SETDB2	13q14	Ocklenburg, Arning et al. (2013a)
PCSK6	15q26.3	Arning et al. (2013); Brandler et al. (2013); Robinson, Hurd, Read, and Crespi (2016); Scerri et al. (2011)
APOE	19q13.2	Bloss, Delis, Salmon, and Bondi (2010)
COMT	22q11.21	Savitz, van der Merwe, Solms, and Ramesar (2007)
AR	Xq12	Arning et al. (2015); Medland et al. (2005)

Table 12.2. Genes associated with language lateralization

Gene name	Location	Study
CCKAR	4p15.2	Ocklenburg, Arning et al. (2013a)
FAT4	4q28.1	Beste, Ocklenburg, von der Hagen, and Di Donato (2016)
KIAA0319	6p22.3	Pinel et al. (2012)
FOXP2	7q31	Ocklenburg, Westerhausen, Hirnstein, and Hugdahl (2013); Pinel et al. (2012)
DCHS1	11p15.4	Beste et al. (2016)
GRIN2B	12p12	Ocklenburg et al. (2011)

RELEVANCE FOR THE GESCHWIND-GALABURDA HYPOTHESIS

The GGH (Geschwind & Galaburda, 1985a, b, c) assumes that intrauterine testosterone levels affect brain structure, which subsequently determines functional lateralization. Thus, of the genes listed in Tables 12.1 and 12.2, the androgen receptor (AR) gene is of interest in the framework of this hypothesis. An important assumption of the GGH (Geschwind & Galaburda, 1985a, b, c) is the idea that the effect of testosterone on brain structure and functional lateralization does not not only depend on the absolute level of free intrauterine testosterone but also on tissue sensitivity to testosterone. The efficacy of the AR gene mediating the actions of testosterone and dihydrotestosterone is one important factor determining individual sensitivity to testosterone. Interestingly, the AR gene comprises a highly polymorphic CAG repeat in exon 1 encoding a glutamine tract in the N-terminal transactivation domain of the protein. A negative linear association between AR function and the CAG-repeat block size has been shown, with longer CAG-repeat blocks linked to less efficient AR function (i.e., decreased testosterone sensitivity in the GGH).

Arning et al. (2015) investigated the role of AR CAG-repeat length and functional lateralization. Mixed-handedness in male participants was significantly associated with longer CAG-repeat blocks. Moreover, female participants homozygous for longer CAG-repeat blocks showed a tendency for stronger left-handedness. These results suggest an association of AR CAG-repeat length and handedness, with longer repeats related to a higher incidence of non–right-handedness. Differences in AR signaling during development may be one of the factors determining individual brain lateralization. Within the framework of the GGH they might contribute to testosterone sensitivity of different types of tissue.

TOWARDS A PARTIAL PLEIOTROPY MODEL FOR THE ONTOGENESIS OF HANDEDNESS AND LANGUAGE LATERALIZATION

No genes that show a significant association with both handedness and language lateralization have been identified (see Tables 12.1 and 12.2), which strongly argues against the notion of a single gene determining both phenotypes. Moreover, the neurophysiological basis of language is different from hand use. Left-lateralized networks in the temporal cortex and the inferior frontal cortex control syntactic processes, whereas less lateralized temporo-frontal networks control semantic processes (Friederici, 2011). In contrast, handwriting, a skill commonly used to assess handedness, has been associated with activation in primary motor, premotor and supplementary motor areas, and superior parietal cortex, as well as thalamus and putamen (Planton, Jucla, Roux, & Démonet, 2013). The distribution of phenotypes also argues against an identical ontogenetic base for handedness and language lateralization. Although the convergence is high for right handers, with only about 5% of right handers being right lateralized for language, the reverse does not hold true; 75% of left handers still show left rather than right lateralization for language. Therefore, we argue that the two traits only show partial pleiotropy, with some shared, but also a substantial amount of independent ontogenetic influences for each of them (see Figure 12.1).

THE MOLECULAR LINK BETWEEN ATYPICAL LATERALIZATION AND SCHIZOPHRENIA

Understanding the ontogenesis of functional hemispheric asymmetries is particularly important in the face of the high prevalence of atypical asymmetries in certain mental and developmental disorders, which suggests that genetic factors contributing to hemispheric asymmetries might also influence the pathogenesis of these disorders. A failure in the lateralization process, which is orchestrated by genes, could be one primary cause of schizophrenia (Crow, 2000). The molecular genetic evidence for a link

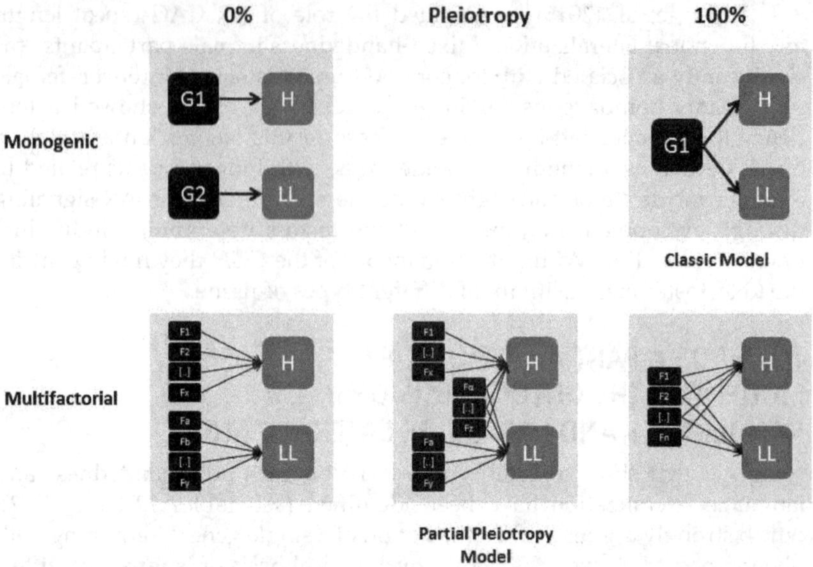

Figure 12.1. Theoretical models for the ontogenesis of handedness and language lateralization. Models differ regarding the amount of pleiotropy between the traits (ranging from 0% to 100%) and the number of genetic and nongenetic factors thought to be relevant (between one and possibly several hundred). (From Ocklenburg, S., Beste, C., Arning, L., Peterburs, J., & Güntürkün, O. [2014]. The ontogenesis of language lateralization and its relation to handedness. *Neuroscience and Biobehavioral Reviews, 43,* 191–198.) (Key: H, handedness; LL, language lateralization; G, genetic influence factor; F, an ontogenetic influence factor that could be genetic, epigenetic, or environmental.)

between laterality and schizophrenia is weak, however. Moreover, there are several other problems with the assumption that schizophrenia and functional lateralization share an identical ontogenetic base. First, schizophrenia is a heterogeneous disorder. The diagnostic criteria according to the *Diagnostic and Statistical Manual of Mental Disorders, Fifth Edition* (*DSM-5;* APA, 2013) require patients to show two or more of the following "A" criteria for a significant amount of time over the course of at least 1 month: delusions, hallucinations, disorganized speech (e.g., frequent derailment), grossly disorganized or catatonic behavior, or negative symptoms (i.e., diminished emotional expression). It seems unlikely that all of these symptoms are driven by the same genetic processes, given that they are likely to originate in different neuronal systems (e.g., auditory verbal hallucinations in the inferior frontal speech regions, catatonic behavior in the motor cortex). In addition, genetic evidence indicates that schizophrenia is not a single disorder but a group of heritable disorders caused by different genotypic networks, leading to distinct clinical symptoms. Arnedo et al. (2015) showed that genetically, there is not one schizophrenia but 17 distinct schizophrenias, with little overlap in involved SNPs and distinct symptoms. Genetic variation seems to affect specific symptoms (e.g., auditory hallucinations) and only indirectly affect the schizophrenia diagnosis per se. Moreover, not all people with schizophrenia show reduced language lateralization. Løberg, Jørgensen, and Hugdahl (2002) reported that young patients without auditory verbal hallucinations showed normal lateralization in the dichotic listening task. Subsequently, Hugdahl et al. (2008) found that whether people with schizophrenia showed atypical language lateralization depended on symptom severity. Patients showed greater reduction of language lateralization with a higher number of positive symptoms. A relation between occurrence of specific symptoms and reduced language lateralization in people with schizophrenia was also confirmed in a meta-analysis of dichotic listening studies by Ocklenburg, Westerhausen, Hirnstein, and Hugdahl (2013). People with schizophrenia showed reduced language lateralization compared with healthy controls. However, the effect size was small ($g = -0.26$). Interestingly, patients with schizophrenia who hallucinated showed a substantially stronger reduction of language lateralization than patients with schizophrenia who did not hallucinate (effect size: $g = -0.45$). These findings highlight the importance of specific symptoms for language lateralization, rather than the schizophrenia diagnosis per se. A paradigm shift needs to occur in which genes are not mapped on schizophrenia as a whole but on discrete schizophrenia symptoms in order to uncover the link between schizophrenia and laterality (Ocklenburg, Güntürkün, Hugdahl, & Hirnstein, 2015; see Figure 12.2).

DYSLEXIA AND ATYPICAL LATERALIZATION

Dyslexia had been linked to atypical lateralization within the context of the GGH (Geschwind & Galaburda, 1985a, b, c). Bishop (2013) suggested, however, that atypical language lateralization might be a consequence, not a cause, of impaired language learning. A symptom-based partial pleiotropy approach may be useful in understanding the causal relationship between atypical lateralization and developmental disorders such as dyslexia. It becomes evident that there is at least some amount of partial pleiotropy when comparing genes related to functional lateralization with those implicated in dyslexia (e.g., *DCDC2*, *ROBO1*, *DYX1C1*, *KIAA0319*) (Paracchini, Scerri, & Monaco, 2007). For example, *KIAA0319* has been implicated in both dyslexia and language lateralization. Moreover, there is also evidence for a symptom-based relation of psychopathology and laterality. Helland, Asbjørnsen, Hushovd, and Hugdahl (2008) reported that a group of people with dyslexia who showed few symptoms and responded well to educational support showed the same dichotic listening performance as healthy controls. In contrast, a second group of people with dyslexia who did not respond to educational support and showed stronger symptoms also showed reduced language lateralization. Thus, taken together, a symptom-based partial pleiotropy approach might be promising to atypical lateralization in dyslexia.

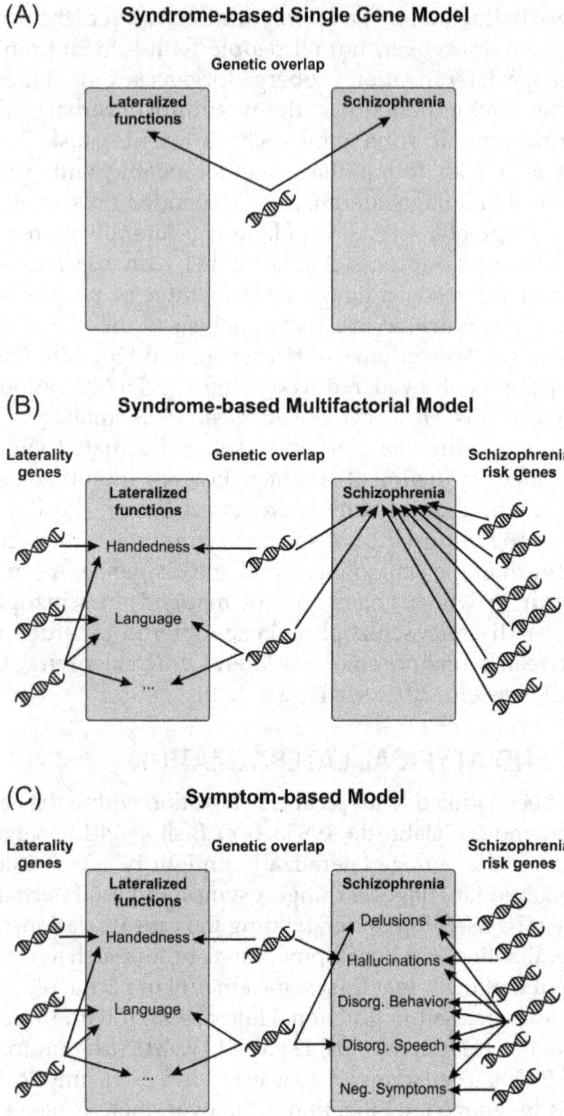

Figure 12.2. Possible models for the genetic association between atypical lateralization and schizophrenia. A) The simple syndrome-based single-gene model assumes that both atypical lateralization and schizophrenia are determined by the same gene. B) Because a considerable body of evidence suggests that both schizophrenia and different forms of lateralization are multifactorial traits, a syndrome-based multifactorial model assuming a multitude of partly overlapping genetic influences for both schizophrenia risk and lateralization is probably a more realistic depiction of the association between the two traits. C) Because specific symptoms have been associated with reduced language lateralization, a symptom-based partial pleiotropy model, with differential genetic (or nongenetic) influences on different schizophrenia-related symptoms, is probably the most realistic model. (From Ocklenburg, S., Güntürkün, O., Hugdahl, K., & Hirnstein, M. [2015]. Laterality and mental disorders in the postgenomic age: A closer look at schizophrenia and language lateralization. *Neuroscience and Biobehavioral Reviews, 59,* 100–110.)

CONCLUSION

The evidence outlined in this chapter suggests that different forms of functional lateralization are complex traits determined by polygenic influences partly overlapping with schizophrenia risk genes. Based on these considerations, a symptom-based partial pleiotropy approach might be helpful to understand the relation of lateralization and psychiatric or developmental disorders on an ontogenetic level.

Future Directions: Toward a Geschwind-Galaburda Hypothesis 2.0

The GGH is a complex model, but in its simplest form it assumes that differences in intrauterine testosterone levels and sensitivity to testosterone affect brain structure, functional lateralization, immune parameters, and LDs. The human genome had not been sequenced and many molecular techniques available today were unattainable when the GGH was presented in the mid-1980s. Many theoretical constructs from this era have not withstood the test of time because considerations based on statistical observations on phenotypes were not actually backed by molecular evidence once that became available. The core assumption of the GGH concerning lateralization—it is determined to a large amount by nongenetic and to a smaller amount by genetic factors—is largely supported by research previously described (Medland et al., 2009). Also, there is evidence for a molecular base of some of the more specific assumptions of the GGH, such as the relation of AR CAG repeats, testosterone, and handedness. Lateralization research is in a theory limbo as it has become evident that phenotype-driven single-gene models are not able to explain the molecular data, and no one has come up with a convincing multifactorial model. The core assumptions of the GGH provide a great foundation to develop such a model. Not all of the assumptions made 30 years ago will likely be supported by genetic or epigenetic data, and new assumptions might need to be added to the model. Careful integration of cutting-edge neurogenetic evidence into the core constructs of the GGH, however, could result in a GGH 2.0 that will hopefully inspire as many researchers over the next 30 years as the original has done over the last three decades.

REFERENCES

American Psychiatric Association. (2013). *Diagnostic and statistical manual of mental disorders, fifth edition (DSM-5)*. Washington, DC: Author.

Annett, M. (1996). In defence of the right shift theory. *Perceptual and Motor Skills, 82,* 115–137.

Armour, J. A., Davison, A., & McManus, I. C. (2014). Genome-wide association study of handedness excludes simple genetic models. *Heredity (Edinburgh), 112,* 221–225.

Arnedo, J., Svrakic, D. M., Del Val, C., Romero-Zaliz, R., Hernández-Cuervo, H., Molecular Genetics of Schizophrenia Consortium,…Zwir, I. (2015). Uncovering the hidden risk architecture of the schizophrenias: Confirmation in three independent genome-wide association studies. *American Journal of Psychiatry, 172,* 139–153.

Arning. L., Ocklenburg, S., Schulz, S., Ness, V., Gerding, W. M., Hengstler, J. G.,…Beste, C. (2013). PCSK6 VNTR polymorphism is associated with degree of handedness but not direction of handedness. *PLoS One, 8,* e67251.

Arning, L., Ocklenburg, S., Schulz, S., Ness, V., Gerding, W. M., Hengstler, J. G.,…Beste, C. (2015). Handedness and the X chromosome: The role of androgen receptor CAG-repeat length. *Scientific Reports, 5,* 8325.

Beste, C., Ocklenburg, S., von der Hagen, M., & Di Donato, N. (2016). Mammalian cadherins DCHS1-FAT4 affect functional cerebral architecture. *Brain Structure and Function, 221,* 2487–2491.

Bishop, D. V. (2013). Cerebral asymmetry and language development: cause, correlate, or consequence? *Science, 340,* 1230531.

Bloss, C. S., Delis, D. C., Salmon, D. P., & Bondi, M. W. (2010). APOE genotype is associated with left-handedness and visuospatial skills in children. *Neurobiology of Aging, 31,* 787–795.

Brandler, W. M., Morris, A. P., Evans, D. M., Scerri, T. S., Kemp, J. P., Timpson, N. J.,… Paracchini, S. (2013). Common variants in left/right asymmetry genes and pathways are associated with relative hand skill. *PLoS Genetics, 9,* e1003751.

Crow, T. J. (2000). Schizophrenia as the price that homo sapiens pays for language: A resolution of the central paradox in the origin of the species. *Brain Research Brain Research Reviews, 31,* 118–129.

Eriksson, N., Macpherson, J. M., Tung, J. Y., Hon, L. S., Naughton, B., Saxonov, S.,…Mountain, J. (2010). Web-based, participant-driven studies yield novel genetic associations for common traits. *PLoS Genetics, 6,* e1000993.

Francks, C., Maegawa, S., Laurén, J., Abrahams, B. S., Velayos-Baeza, A., Medland, S. E.,… Monaco, A. P. (2007). LRRTM1 on chromosome 2p12 is a maternally suppressed gene that is associated paternally with handedness and schizophrenia. *Molecular Psychiatry, 12,* 1129–1139.

Friederici, A. D. (2011). The brain basis of language processing: from structure to function. *Physiology Reviews, 91,* 1357–1392.

Geschwind, N., & Galaburda, A. M. (1985a). Cerebral lateralization. Biological mechanisms, associations, and pathology: I. A hypothesis and a program for research. *Archives of Neurology, 42*(5), 428–459.

Geschwind, N., & Galaburda, A. M. (1985b). Cerebral lateralization. Biological mechanisms, associations, and pathology: II. A hypothesis and a program for research. *Archives of Neurology, 42*(6), 521–552.

Geschwind, N., & Galaburda, A. M. (1985c). Cerebral lateralization. Biological mechanisms, associations, and pathology: III. A hypothesis and a program for research. *Archives of Neurology, 42*(7), 634–654.

Helland, T., Asbjørnsen, A. E., Hushovd, A. E., & Hugdahl, K. (2008). Dichotic listening and school performance in dyslexia. *Dyslexia, 14,* 42–53.

Hugdahl, K., Løberg, E. M., Jørgensen, H. A., Lundervold, A., Lund, A., Green, M. F., & Rund, B. (2008). Left hemisphere lateralisation of auditory hallucinations in schizophrenia: A dichotic listening study. *Cognitive Neuropsychiatry, 13,* 166–179.

Kavaklioglu, T., Ajmal, M., Hameed, A., & Francks, C. (2016). Whole exome sequencing for handedness in a large and highly consanguineous family. *Neuropsychologia, 93,* 342–349.

Li, Q., Bian, S., Liu, B., Hong, J., Toth, M., & Sun, T. (2013). Establishing brain functional laterality in adult mice through unilateral gene manipulation in the embryonic cortex. *Cell Research, 23,* 1147–1149.

Løberg, E. M., Jørgensen, H. A., & Hugdahl, K. (2002). Functional brain asymmetry and attentional modulation in young and stabilised schizophrenic patients: A dichotic listening study. *Psychiatry Research, 109,* 281–287.

McManus, I. C., Davison, A., & Armour, J. A. (2013). Multilocus genetic models of handedness closely resemble single-locus models in explaining family data and are compatible with genome-wide association studies. *Annals of the New York Academy of Sciences, 1288,* 48–58.

Medland, S. E., Duffy, D. L., Spurdle, A. B., Wright, M. J., Geffen, G. M., Montgomery, G. W., & Martin, N. G. (2005). Opposite effects of androgen receptor CAG repeat length on increased risk of left-handedness in males and females. *Behavior Genetics, 35,* 735–744.

Medland, S. E., Duffy, D. L., Wright, M. J., Geffen, G. M., Hay, D. A., Levy, F.,…Boomsma, D. I. (2009). Genetic influences on handedness: Data from 25,732 Australian and Dutch twin families. *Neuropsychologia, 47,* 330–337.

Ocklenburg, S., Arning, L., Gerding, W. M., Epplen, J. T., Güntürkün, O., & Beste, C. (2013a). Cholecystokinin A receptor (CCKAR) gene variation is associated with language lateralization. *PLoS One, 8,* e53643.

Ocklenburg, S., Arning, L., Gerding, W. M., Epplen, J. T., Güntürkün, O., & Beste, C. (2013b). FOXP2 variation modulates functional hemispheric asymmetries for speech perception. *Brain and Language, 126,* 279–284.

Ocklenburg, S., Arning, L., Gerding, W. M., Hengstler, J. G., Epplen, J. T., Güntürkün, O.,…Akkad, D. A. (2016). Left-right axis differentiation and functional lateralization: A haplotype in the methyltransferase encoding gene SETDB2 might mediate handedness in healthy adults. *Molecular Neurobiology,* 53, 6355–6361.

Ocklenburg, S., Arning, L., Hahn, C., Gerding, W. M., Epplen, J. T., Güntürkün, O., & Beste, C. (2011). Variation in the NMDA receptor 2B subunit gene GRIN2B is associated with differential language lateralization. *Behavioural Brain Research, 225,* 284–289.

Ocklenburg, S., Beste, C., Arning, L., Peterburs, J., & Güntürkün, O. (2014). The ontogenesis of language lateralization and its relation to handedness. *Neuroscience and Biobehavioral Reviews, 43,* 191–198.

Ocklenburg, S., Güntürkün, O., Hugdahl, K., & Hirnstein, M. (2015). Laterality and mental disorders in the postgenomic age: A closer look at schizophrenia and language lateralization. *Neuroscience and Biobehavioral Reviews, 59,* 100–110.

Ocklenburg, S., Westerhausen, R., Hirnstein, M., & Hugdahl, K. (2013). Auditory hallucinations and reduced language lateralization in schizophrenia: A meta-analysis of dichotic listening studies. *Journal of the International Neuropsychological Society, 19,* 410–418.

Paracchini, S., Scerri, T., & Monaco, A. P. (2007). The genetic lexicon of dyslexia. *Annual Review of Genomics and Human Genetics, 8,* 57–79.

Pinel, P., Fauchereau, F., Moreno, A., Barbot, A., Lathrop, M., Zelenika, D.,…Dehaene, S. (2012). Genetic variants of FOXP2 and KIAA0319/TTRAP/THEM2 locus are associated with altered brain activation in distinct language-related regions. *Journal of Neuroscience, 32,* 817–825.

Planton, S., Jucla, M., Roux, F. E., & Démonet, J. F. (2013). The "handwriting brain": A meta-analysis of neuroimaging studies of motor versus orthographic processes. *Cortex, 49,* 2772–2787.

Robinson, K. J., Hurd, P. L., Read, S., & Crespi, B. J. (2016). The PCSK6 gene is associated with handedness, the autism spectrum, and magical ideation in a non-clinical population. *Neuropsychologia, 84,* 205–212.

Savitz, J., van der Merwe, L., Solms, M., & Ramesar, R. (2007). Lateralization of hand skill in bipolar affective disorder. *Genes, Brain and Behavior, 6,* 698–705.

Scerri, T. S., Brandler, W. M., Paracchini, S., Morris, A. P., Ring, S. M., Richardson, A. J., & Monaco, A. P. (2011). PCSK6 is associated with handedness in individuals with dyslexia. *Human Molecular Genetics, 20,* 608–614.

Somers, M., Ophoff, R. A., Aukes, M. F., Cantor, R. M., Boks, M. P., Dauwan, M.,…Sommer, I. E. (2015). Linkage analysis in a Dutch population isolate shows no major gene for left-handedness or atypical language lateralization. *Journal of Neuroscience, 35,* 8730–8736.

INTEGRATIVE SUMMARY 3

From Genes to Brains

Silvia Paracchini

SUMMARY

Chapters 11 and 12 address the molecular mechanisms at the basis of laterality and discuss the recent advances in the field of genetics. Chapter 11 explores the challenges of the field and the underappreciation of the complexity of the human genome. In fact, we have started to grasp that the genetics of traits such as handedness are more complex than what was anticipated decades ago. Chapter 12 describes some of the specific genes that have been identified so far, but it is clear that they represent only a small fraction of the biological component controlling the establishment of laterality and handedness. I summarize the two chapters in this section and explore how they complement each other and relate to the Geschwind-Galaburda hypothesis (GGH; Geschwind & Galaburda, 1985a, b, c). In particular, we have learned that simple genetic models are unlikely to control complex processes. The advances in sequencing technology, which are now allowing us to interrogate the genome at unprecedented resolution, are likely to play a major role in advancing our understanding of laterality in the near future.

SECTION COMMENTARY

The two chapters of this section discuss the molecular mechanisms at the basis of lateralization from what can be considered the most upstream element on the chain of events—the gene, or better said, the genes. Genetic elements have been central in the development of the GGH. The field of human genetics has developed at an extremely fast rate since the mid-2000s as a result of the completion of the Human Genome Project (Lander, 2011) as well as advances in technology for generating data. The latter has probably been the main driving force that has changed the landscape of this discipline. In fact, generating genotype data, which used to be laborious and expensive, has become easy, and the real challenge lies now in making sense of massive amounts of data generated at relative low cost with unprecedented high throughput. In light of these developments, these two chapters, as well as this entire volume, offer an opportunity to reflect on how advances in this field have influenced our understanding of the role of asymmetries in neurodevelopment and in neurodevelopmental disorders (NDs).

Silvia Paracchini is a Royal Society University Research Fellow.

Grigorenko (Chapter 11) takes us through different layers of complexity that are typically encountered for genetic studies. The first level of complexity is represented by different classes of genetic variants, ranging from single nucleotide polymorphisms (SNPs), which typically do not affect the direct product of a gene (i.e., a protein), to large genetic rearrangements, such as copy number variants (CNVs), which might completely remove or add entire genes. The effect of a genetic variant on the function of a gene cannot, however, be predicted solely on the length and location of its sequence. For example, gene function can be modified by epigenetic factors, which can be considered a second layer of complexity. Epigenetic factors are typically represented by chemical modifications added to DNA molecules that do not change the sequence of a gene but might regulate when and where genes are activated or silenced. Epigenetic factors are highly dynamic and might have variable effects across different tissues and at different times of development. This dynamism makes epigenetics susceptible to environmental effects and can provide a very attractive platform to directly test how external factors (e.g., chemicals, diet, stressors) might affect biological processes. This very same dynamism presents significant complications, however, such as difficulties in replicating potential effects or gaining access to the correct tissue. In the case of neurodevelopmental traits, we would ideally need to analyze epigenetic differences across brain tissues from different individuals. These types of tissues are obviously inaccessible and we cannot assume that what can be observed in more accessible tissues such as blood or saliva would be representative of epigenetic changes that affect brain development. The fact that the same epigenetic factors might not have the same downstream effects is a further complication. It has been shown that the same SNPs might be associated with different cognitive traits (Newbury et al., 2011), and even rearrangements normally observed in patients with severe phenotypes such as epilepsy and intellectual disabilities can specifically affect language (Pettigrew et al., 2015). These observations suggest that genes work within networks and, therefore, their effects might depend on interactions with other genetic factors.

The difficulty in disentangling individual gene effects is also a consequence of our inability to directly measure gene effects, introducing another important challenge for conducting genetic studies of NDs—the lack of measurable biological markers. One way to get around this problem is represented by the so-called endophenotypes. The principle of endophenotypes is to use something that we can measure, such as neuroimaging measures or cognitive/behavioral tests, instead of simply relying on disorder diagnoses for genetic analyses. Grigorenko (Chapter 11) rightly explains, however, that it has been challenging to identify reliable associations between genetic markers and, for example, brain imaging data; in fact, this area of research has been affected more than others by the reproducibility crisis featured prominently in current scientific debates (Baker, 2016;

Button et al., 2013). The problem usually stems from the difficulty of collecting these types of detailed measures in large enough numbers of individuals. Reduced power in combination with many variables inevitably leads to false positives. The field is developing solutions to tackle this issue by forming large consortia to facilitate data sharing, which has allowed the analysis of brain asymmetries in more than 15,000 cases (Guadalupe et al., 2016).

Ocklenburg and Peterburs (Chapter 12) describe more directly the role of specific genes identified so far through genetic studies. One certain conclusion is that we can safely exclude the effect of a single gene in driving handedness and language lateralization. Given the large number of individuals for whom hand preference measures and genome-wide SNP data are available, we would have mapped at least one gene associated with left-handedness by now if such a gene existed (Paracchini & Scerri, 2017). This observation takes single-gene theories, which dominated the field for a long time, out of the picture. Instead, things are far more complicated (see Chapter 11). We should treat handedness and lateralization as complex traits and expect multiple genetic factors to be contributing to them. The number of potential candidate genes remains very small, in spite of significant efforts. Conducting genetic studies for language lateralization or cerebral asymmetries is challenging compared with handedness studies because of the difficulty in gathering such types of data. Among the few genes reported as having a role in language lateralization, no obvious overlap exists between the genes reported to be associated with handedness (see Tables 12.1 and 12.2).

It is worth noting, however, that a few candidates for handedness are particularly interesting in the context of the GGH. The androgen receptor (*AR*) gene is located on the X chromosome and is implicated in mediating testosterone effects. A microsatellite marker (i.e., a marker made up of variable numbers of short stretches of DNA repeated in variable numbers across individuals, the CAG triplet in this case) was associated with handedness, with females and males more likely to be left handed if they had a greater or smaller number of repeats, respectively (Medland et al., 2005). Follow-up studies highlighted the importance of identifying mixed handers to dissect the association trends at this locus (Arning et al., 2015; Hampson & Sankar, 2012). How handedness is measured and in which subgroups genetic analysis is conducted have been two crucial elements for the identification of other candidate genes that are also relevant for the GGH. The original observation of reduced asymmetries of the planum temporale in individuals with dyslexia has been consistently replicated across multiple studies, and a role of atypical asymmetries has been consistently described for different NDs, including schizophrenia, bipolar disorder, autism spectrum disorder, and stuttering. For this reason, handedness has been specifically assessed in individuals with NDs beyond the collection of hand preference data. Both the *LRRTM1* and *PCSK6* association with handedness derived from the assessment of handedness as a quantitative measure using the pegboard test (Annett, 1985), which assesses how much better an individual performs with one hand versus the other or, in other words, how strongly he or she is lateralized for handedness. *PCSK6* is the only candidate so far that has shown an association with handedness at a statistically significant level (Brandler et al., 2013). The *PCSK6* (Scerri et al., 2011) together with the *LRRTM1* (Francks et al., 2007) associations were specifically identified in individuals with dyslexia, which is of interest for the GGH. Although it remains difficult to explain these dyslexia-specific effects, the results suggest that the same pathways implicated in handedness are also relevant to dyslexia (Brandler & Paracchini, 2014). Therefore, it is important to assess these quantitative measures of handedness in the general population as well as test whether dyslexia-specific effects are genuine or could be the artefacts of small sample sizes (Paracchini, 2011). Although the data indicate a much more complex picture than what was originally hypothesized, Ocklenburg and Peterburs suggest that these recent advances represent an opportunity to revisit the GGH and provide the field with a novel framework to guide future studies.

CONCLUSIONS

The take-home message from these two chapters is that the contribution of genetic factors to asymmetries and their role in disorders is much more complex than could have been hypothesized 30 years ago. The knowledge generated from genetic studies is demonstrating that we cannot expect easy formulas predicting which specific neurodevelopmental traits can be uniquely traced back to a few genes. This has been clearly demonstrated by the failure of large genome-wide association studies to identify genes associated with left-handedness that would fit single-gene theories. The field of neurodevelopmental traits presents particular challenges on how to measure the effects of genes. For example, the use of a quantitative measure of handedness versus the more accessible hand preference has been key to identifying some interesting genes. Therefore, rather than simply looking left versus right, we need to consider the best measures that might pick up gene effects on lateralization and how these measures can be collected in large enough numbers of individuals. International cooperation to achieve large sample sizes is part of the answer but probably not enough. Communication across disciplines such as cognitive neuroscience and human genetics is going to be vital to ensure that meaningful genetic studies will identify candidate genes that might lead to a better understanding of what biological pathways contribute to asymmetries and how these are relevant to disorders.

REFERENCES

Annett, M. (1985). *Left, right, hand and brain: The right shift theory.* London, England: Psychology Press.

Arning, L., Ocklenburg, S., Schulz, S., Ness, V., Gerding, W. M., Hengstler, J. G.,... Oldfield, R. C. (2015). Handedness and the X chromosome: The role of androgen receptor CAG-repeat length. *Scientific Reports, 5,* 8325. http://doi.org/10.1038/srep08325

Baker, M. (2016). 1,500 scientists lift the lid on reproducibility. *Nature, 533*(7604), 452–454. http://doi.org/10.1038/533452a

Brandler, W. M., Morris, A. P., Evans, D. M., Scerri, T. S., Kemp, J. P., Timpson, N. J.,… Paracchini, S. (2013). Common variants in left/right asymmetry genes and pathways are associated with relative hand skill. *PLoS Genetics, 9*(9), e1003751. http://doi.org/10.1371/journal.pgen.1003751

Brandler, W. M., & Paracchini, S. (2014). The genetic relationship between handedness and neurodevelopmental disorders. *Trends in Molecular Medicine, 20*(2), 83–90. doi: 10.1016/j.molmed.2013.10.008

Button, K. S., Ioannidis, J. P. A., Mokrysz, C., Nosek, B. A., Flint, J., Robinson, E. S. J., & Munafò, M. R. (2013). Power failure: Why small sample size undermines the reliability of neuroscience. *Nature Reviews Neuroscience, 14*(5), 365–376. http://doi.org/10.1038/nrn3475

Francks, C., Maegawa, S., Lauren, J., Abrahams, B. S., Velayos-Baeza, A., Medland, S. E.,… Monaco, A. P. (2007). LRRTM1 on chromosome 2p12 is a maternally suppressed gene that is associated paternally with handedness and schizophrenia. *Molecular Psychiatry, 12*(12), 1057, 1129–1139. Retrieved from http://www.ncbi.nlm.nih.gov/entrez/query.fcgi?cmd=Retrieve&db=PubMed&dopt=Citation&list_uids=17667961

Geschwind, N., & Galaburda, A. M. (1985a). Cerebral lateralization. Biological mechanisms, associations, and pathology: I. A hypothesis and a program for research. *Archives of Neurology, 42*(5), 428–459.

Geschwind, N., & Galaburda, A. M. (1985b). Cerebral lateralization. Biological mechanisms, associations, and pathology: II. A hypothesis and a program for research. *Archives of Neurology, 42*(6), 521–552.

Geschwind, N., & Galaburda, A. M. (1985c). Cerebral lateralization. Biological mechanisms, associations, and pathology: III. A hypothesis and a program for research. *Archives of Neurology, 42*(7), 634–654.

Guadalupe, T., Mathias, S. R., vanErp, T. G. M., Whelan, C. D., Zwiers, M. P., Abe, Y.,… Francks, C. (2016). Human subcortical brain asymmetries in 15,847 people worldwide reveal effects of age and sex. *Brain Imaging and Behavior,* 1–18. http://doi.org/10.1007/s11682-016-9629-z

Hampson, E., & Sankar, J. S. (2012). Hand preference in humans is associated with testosterone levels and androgen receptor gene polymorphism. *Neuropsychologia, 50*(8), 2018–2025. http://doi.org/10.1016/j.neuropsychologia.2012.04.027

Lander, E. S. (2011). Initial impact of the sequencing of the human genome. *Nature, 470*(7333), 187–197. http://doi.org/10.1038/nature09792

Medland, S. E., Duffy, D. L., Spurdle, A. B., Wright, M. J., Geffen, G. M., Montgomery, G. W., & Martin, N. G. (2005). Opposite effects of androgen receptor CAG repeat length on increased risk of left-handedness in males and females. *Behavior Genetics, 35*(6), 735–744. http://doi.org/10.1007/s10519-005-6187-3

Newbury, D. F., Paracchini, S., Scerri, T. S., Winchester, L., Addis, L., Richardson, A. J.,… Monaco, A. P. (2011). Investigation of dyslexia and SLI risk variants in reading- and language-impaired subjects. *Behavior Genetics, 41*(1), 90–104.

Paracchini, S. (2011). Dissection of genetic associations with language-related traits in population-based cohorts. Journal of Neurodevelopmental Disorders, 3(4), 365–373. http://doi.org/10.1007/s11689-011-9091-6

Paracchini, S., & Scerri, T. (2017). Genetics of human handedness and laterality. In L. J. Rogers & G. Vallortigara (Eds.), *Lateralized brain functions* (pp. 523–552). New York, NY: Springer.

Pettigrew, K. A., Reeves, E., Leavett, R., Hayiou-Thomas, M. E., Sharma, A., Simpson, N. H.,…Paracchini, S. (2015). Copy number variation screen identifies a rare de novo deletion at chromosome 15q13.1-13.3 in a child with language impairment. *PLoS One, 10*(8), e0134997.

Scerri, T. S., Brandler, W. M., Paracchini, S., Morris, A. P., Ring, S. M., Richardson, A. J.,… Monaco, A. P. (2011). PCSK6 is associated with handedness in individuals with dyslexia. *Human Molecular Genetics, 20*(3), 608–614.

SECTION V

Reading and Dyslexia

From Genes to Behavior

CHAPTER 13

Genetics of Specific Reading Disability

The State of Affairs

Sergey A. Kornilov and Elena L. Grigorenko

SUMMARY

Recent advances in the fields of genetics and genomics of complex traits gave rise to a new wave of studies of genetic etiology of specific reading disability (SRD) and reading skills. These studies form three major branches: 1) genome-wide association studies (GWAS) that focus on the role of common genetic variants in the etiology of SRD, 2) studies that use neuroimaging measures as phenotypes, and 3) studies that rely on next-generation DNA sequencing to examine the role of rare genetic variants in SRD. Rooted in the assumption that a large number of common genetic variants might regulate the typical development of reading, a number of GWAS have uncovered new suggestive candidate SRD/reading genes. These genes and variants, however, do not account for the majority of the heritability of reading and SRD. Two solutions to this problem of missing heritability have been proposed. First, the field has been slowly adopting the use of neuroimaging measures (both functional and structural) in addition to behavior measures of reading to study the genetics of SRD. Second, the field started looking more closely at the complementary role of rare but impactful genetic variants in SRD. Although promising, findings from GWAS, endophenotypic, and DNA sequencing studies of SRD are very preliminary. Uncovering the complexities of the genetic etiology of SRD will require a careful replication, functional annotation, and theoretical integration of the results generated by these three branches of molecular genetic studies of SRD.

INTRODUCTION

SRD is a highly prevalent neurodevelopmental disorder (ND) characterized by difficulties in the acquisition of reading and reading-related skills in the absence of other explanatory factors such as intellectual disability.

The writing of this chapter was supported by grants R01 DC007665 and R21 HD070594 from the U.S. National Institutes of Health and grant 14.Z50.31.0027 from the Government of the Russian Federation. Grantees undertaking such projects are encouraged to express freely their professional judgment. The chapter, therefore, does not necessarily reflect the position or policies of these funding agencies, and no official endorsement should be inferred.

Although the empirical literature on the behavioral manifestations and neurocognitive underpinnings of the disorder is voluminous, relatively little is known about its genetic etiology. Thus, the current understanding of the specific mechanisms that constitute and convey genetic risk for SRD is lacking, despite the long history of successful empirical behavioral genetic investigations that have established that SRD has a strong genetic component. The aim of this chapter is to provide the reader with a general overview of the current state of affairs in the field of the genetics of SRD while focusing on the new generation of studies published since the late 2000s.

The early wave of linkage and targeted association studies identified most of the known SRD candidate susceptibility loci at labeled dyslexia susceptibility regions (from DYX1 to DYX9). Together, they span a large number of genes located in nine chromosomal regions: 1p34-p36 (DYX8), 2p15-p16 (DYX3), 3p14.1-q13 (DYX5), 6p22.2 (DYX2), 6q12-q14.1 (DYX4), 11p15.5 (DYX7), 15q21 (DYX1), 18p11 (DYX6), and Xq27 (DYX9). To date, seven candidate genes located within these loci have received support from at least two independent molecular studies: *DYX1C1* at DYX1, *DCDC2* and *KIAA0319* at DYX2, *MRPLI9* and *C2ORF3* at DYX3, *ROBO1* at DYX5, and *KIAA0319L* at DYX8. Several reviews of these studies have been published, and we encourage the reader to refer to them (e.g., Carrion-Castillo, Franke, & Fisher, 2013; Skiba, Landi, Wagner, & Grigorenko, 2011). Given space considerations, we limit this review to the new generation of molecular genetic studies of SRD—studies that used neuroimaging endophenotypes or capitalized on the development of high-throughput genotyping (i.e., single nucleotide polymorphism [SNP] microarrays), copy number variant (CNV) detection, and next-generation DNA sequencing technologies.

As previously noted, multiple twin (Grigorenko, 2004), family (e.g., Wolff & Melngailis, 1994), and adoption (e.g., Brooks, Fulker, & DeFries, 1990) studies have incontrovertibly established that SRD has a strong genetic component and displays significant familiality and heritability. Although phenotypic manifestations of SRD are complex, multifaceted, and heterogeneous, heritability estimates of various facets of SRD hover at 60% on average (Grigorenko, 2004), although there are additional sources of variation in these estimates, such as age (Byrne et al., 2009), ethnicity (Grigorenko, Ngorosho, Jukes, & Bundy, 2006), socioeconomic status and parental education (Friend, DeFries, & Olson, 2008), and quality of classroom environment (Taylor, Roehrig, Soden Hensler, Connor, & Schatschneider, 2010). Similarly, the estimated risk (for holistic diagnoses as well as phenotypic facets) for first-degree relatives of people with SRD is about two to three times higher than that of the general population (Schulte-Körne, Deimel, Muller, Gutenbrunner, & Remschmidt, 1996). Finally, the risk to adopted individuals is associated with the reading performance of their biological but not their adoptive relatives (Wadsworth, Corley, Hewitt, & DeFries, 2001). Although the genetic underpinning of

SRD is quite apparent, the mode of inheritance or overall genetic architecture of the disability and the specific genetic mechanisms that form that underpinning have been substantially less clear and rather controversial. The field is dominated by two major models of the overall genetic architecture of SRD, which is the case for other complex disorders. These models differ fundamentally in their specification of the disability and their interpretation of normal variation in reading performance.

THE CURRENT STATE OF GENETICS AND SPECIFIC READING DISABILITY

Up to now, the field of genetic studies of SRD has been subjugated by the common disorder-common variant (CDCV) hypothesis, which posits that SRD arises on a polygenic background of inheritance of multiple common genetic risk variants that individually are characterized by small effect sizes but collectively surpass a certain liability threshold above which the disability is manifested. It is also assumed that some of these risk variants are general to all facets of SRD and, perhaps, to SRD and other learning disabilities (Plomin & Kovas, 2005), whereas others are SRD specific and even facet specific (Naples, Chang, Katz, & Grigorenko, 2009). These partial overlaps of risk variants can explain many of the current findings, such as the substantial, but far from perfect, genetic correlations between different facets of SRD; the differential heritability estimates for different facets and their fluctuations when other sources of variance are considered; and the flagging of numerous candidate genomic regions harboring relevant genes and numerous genes as candidate genes for SRD (Peterson & Pennington, 2012). Finally, the CDCV hypothesis assumes the continuity between typical and atypical levels of development of complex traits because common risk variants are present in the general population at levels below the liability threshold, and this presence guarantees a continuity of reading performance.

The CDCV hypothesis is at the conceptual and methodological foundation of GWAS, which interrogate multiple (e.g., from several hundred thousand to several million) SNP markers spaced evenly across the genome in large cohorts comprised either of unrelated individuals (population based) or a multitude of familial structures (family based). The goal of GWAS is to identify genetic variants that are either causal with respect to the etiology of the disorder or are in linkage disequilibrium (i.e., correlated) with causal variants because each marker is able to tag and act as a proxy for nearby variants even if they were not directly genotyped.

Several GWAS of SRD defined through a categorical diagnostic label (SRD *vs.* typically developing) or through continuous reading phenotypes (e.g., single-word reading, spelling) have been published, producing a complex and heterogeneous landscape of findings. Luciano et al. (2013) used two large population samples of Australian twins and siblings (n = 1,177 individuals ages 12–25 years) and singletons from the Avon Longitudinal Study of Parents and Children (ALSPAC; *n* = 5,472 individuals ages 8–9 years; Golding & ALSPAC Study Team, 2004) and performed a GWAS of reading and language traits, interrogating more than 610,000 SNP markers genotyped using the Illumina 610k Quad BeadChip microarray platform (imputed to 2,373,249 analyzed SNPs total). Although no SNPs survived corrections for multiple testing, suggestive marker-based and several significant gene-based associations have been identified for three quantitative phenotypes (pseudoword repetition, reading, and spelling skills), implicating genes *DAZAP1* (DAZ-associated protein 1; located on chromosome 19p13.3), *CD2L1* (cell division cycle 2-like 1; 1p36), *CDC2L2* (cell division cycle 2-like protein kinase 2; 1p36.33), and *RCAN3* (RCAN family member 3; 1p36.11), and one pseudogene (*ABCC13;* ATP-binding cassette subfamily C member 13; 21q11.2) in reading. These novel findings are promising given the involvement of the new candidates in a variety of developmental processes and associated molecular pathways, including cell growth and cycle progression, and Ca^{2+}-mediated signaling. Interestingly, none of these new candidates have previously been implicated in language and reading, and, simultaneously, early SRD candidates such as *KIAA0319, DCDC2*, and *DYX1C1* did not show significant associations in this study.

Eicher et al. (2013) performed a GWAS of reading and language in a sample of SRD (*n* = 353), language impaired (LI; *n* = 163), SRD+LI (*n* = 174), and typically developing children (*n* = 4,117) identified and selected from the ALSPAC sample based on available continuous measures of reading and language functioning. The authors used the same genotyping panel but a slightly different approach to the conceptualization of the sample and data analysis and identified two novel genes as possibly conferring a risk for the comorbid SRD+LI—*ZNF385D* (zinc finger protein 385D; 3p24.3) and *COL4A2* (collagen type IV alpha 2; 13q34) and one gene associated with LI—*NDST4* (N-deacetylase/N-sulfotransferase 4; 4q26). The authors followed up on their findings and used the data from the Pediatric Imaging, Neurocognition, and Genetics study (PING; *n* = 440 children; Jernigan et al., 2016), genotyped on the Illumina Human660W-Quad BeadChip platform. They successfully replicated the association signal for *ZNF385D,* although a receptive vocabulary measure was used as the phenotype in this case.

Gialluisi et al. (2014) performed a GWAS meta-analysis, pooling data from three separate cohorts (UK Reading Disability, SLI Consortium, and Colorado Learning Disabilities Research Centre) for a total effective sample size of *n* = 1,862. This study identified several novel, albeit suggestive, associations between the first principal component of a number of language and reading measures and two SNPs—rs59197085 located in *CCDC136/FLNC* (filamin C; 7q32.1) and rs5995177 located in *RBFOX2* (RNA-binding protein, Fox-1 homolog 2; 22q12.3). The *FLNC* association was further replicated in a large (*n* = 6,434) independent cohort.

Six other studies (Eicher et al., 2016; Gialluisi, Guadalupe, Francks, & Fisher, in press; Jasińska et al., 2016; Landi et al., 2013; Roeske et al., 2011; Skeide et al., 2015) deserve special mention because of their focus on endophenotypes—intermediate phenotypes between the levels of observable phenotypic deficits and their genetic etiology. Endophenotypes have been argued to provide a more sensitive and detailed characterization of representational and processing deficits at the level likely more proximal to the etiology of the disorder (i.e., neurocognitive for NDs such as SRD) than end-point behavioral performance measures or umbrella diagnostic categories (Cannon & Keller, 2006), but they have been severely underutilized in studies of NDs.

Roeske et al. (2011) performed a GWAS of the electrophysiological component called *mismatch negativity* (MMN) in response to speech stimuli in a small sample of $n = 200$ individuals with SRD, genotyped on the Illumina HumanHap300k BeadChip platform. They found a suggestive association between the amplitude of the late MMN and rs4234898 located on chromosome 4q32.1. The association has been replicated in an independent sample of $n = 186$ children with SRD and reached genome-wide significance in the combined analysis of the two samples. Although this SNP is located in the region that does not contain protein-coding genes, it nevertheless shows associations with neuronal expression levels of *SLC2A3* (solute carrier family 2 member 3; 12p13.31), a gene that codes for the predominant facilitative transporter of glucose in neurons. This study provides an example of a successful, albeit modest in scope, application of the endophenotype idea to the investigation of molecular genetic bases of SRD and provides tentative support for the roles of speech MMN and glucose metabolism (heavily involved in synapse formation and branching) in the etiology of SRD.

Three published targeted association studies of candidate SRD genes specifically focused on neuroimaging endophenotypes while examining early SRD candidates (Eicher et al., 2016) as well as the new GWAS-related ones (Gialluisi et al., in press; Skeide et al., 2015). Thus, Eicher et al. examined a set of SNPs located within the DYX2 and DYX3 loci, using cortical volume, cortical thickness, and fractional anisotropy as phenotypes for the association analysis of the PING dataset ($n = 332$). They found that markers in *KIA0319* and *FAM65B* were associated with cortical thickness in the left orbitofrontal region and global fractional anisotropy. Several other tentative associations have been established, and several previous findings were not replicated, including the previously reported association between DYX3 and hippocampal measures.

Gialluisi et al. (in press) performed a targeted association study focusing on SNPs located within *FLNC* and *RBFOX2*—genes identified in a behavioral GWAS of SRD in an earlier study (Gialluisi et al., 2014). Using MRI-based neuroanatomical measures (gray matter surface area and thickness) from five brain regions implicated in language and reading obtained from $n = 1{,}275$ individuals, they found that the minor allele of rs5995177 located in *RBFOX2* was significantly associated with reduced cortical thickness in a multivariate analysis, showing the same direction of effect as the behavioral GWAS. Similarly, Skeide et al. (2015) examined rs11100040, an SNP reported to affect *SLC2A3* expression, in a small sample of $n = 32$ children, focusing on patterns of functional as well as structural connectivity in the brain. They found that rs11100040 was associated with resting-state functional connectivity of left fronto-temporal areas involved in phonological processing as well as with the fractional anisotropy of the arcuate fasciculus connecting these areas.

Several functional neuroimaging studies suggested that other genes previously implicated in neural development and cognition might be associated with reading and SRD. Thus, Landi et al. (2013) found that rs4680, a well-characterized common variant in *COMT* (catechol-o-methyltransferase; 22q11.21), a gene coding for one of the enzymes that regulate dopamine turnover, was associated with patterns of functional brain activation during reading and behavioral indices of phonological processing in a sample of $n = 86$ children. In this study, children who carried one or two copies of the minor allele at that locus (linked to higher levels of prefrontal dopamine) showed superior performance and patterns of the functional brain activation more characteristic of better readers compared with children who carried two major alleles. Similarly, Jasińska et al. (2016) focused on rs6265, a common polymorphism in the *BDNF* (brain-derived neurotrophic factor; 11p14.1) gene known to regulate the secretion of the BDNF protein involved in neuronal development and brain plasticity. In this study of $n = 81$ children, carriers of the minor allele at this locus (linked to lower levels of BDNF in the brain) showed poorer behavioral performance on memory-taxing phonological tasks and functional overactivation of brain areas involved in reading compared with children who carried two copies of the major allele. The last six studies point to the beginning of a new era of interdisciplinary research into the neurobiological underpinnings of SRD and suggest the need for a careful consideration of neuroimaging endophenotypes as well as the set of interrogated candidates in future studies.

Finally, we would like to mention a GWAS directly relevant to the topic of the 2016 GGH symposium held by The Dyslexia Foundation. Brandler et al. (2013) focused on a quantitative measure of relative hand skill in a meta-analysis of several samples of individuals with SRD ($n = 728$) genotyped on the Illumina 550K BeadChip platform. Relative hand skill, an endophenotype directly relevant to the GGH (Geschwind & Galaburda, 1985a, b, c), was found to be associated with rs7182874, located in *PCSK6* (proprotein convertase subtilisin/kexin type 6; 15q26.3), in this sample but not in a larger sample from the general population ($n = 2,666$), suggesting that the association is SRD specific. The authors used gene-set enrichment analysis and demonstrated a significant overrepresentation of SNPs located within human orthologs of genes that regulate left/right asymmetry knockout phenotypes in mice. This study provided intriguing if indirect evidence in support of the role of genes involved in left/right pattern formation in handedness-relevant phenotypes in SRD.

As previously noted, GWAS rely on the CDCV hypothesis and the ability of existing genotyping platforms to capture the relevant common variation in the genome through a large set of tag SNPs. The GWAS previously mentioned had a modest statistical power compared with those performed for other human phenotypes, including psychiatric disorders, and have identified a multitude of tentative novel associations, implicating new genes in the etiology of SRD. These novel associations are very small with respect to their effect sizes, accounting for under 1% of the phenotypic variance, thus explaining under 1.6% of the SRD heritability estimates (assuming $M = 60\%$) from behavioral genetic studies. This situation is parallel to the general findings from the GWAS era of molecular genetic research on other complex human traits and common disorders; when pooled together, disease- and trait-associated SNPs typically explain well under 10% of phenotypic variance in complex human traits (Visscher, Brown, McCarthy, & Yang, 2012). Because this only accounts for a minor fraction of the estimated heritability, the remaining unexplained portion of heritability has been called *missing heritability* (Manolio et al., 2009).

CONCLUSION

Several factors have been proposed to account for the missing heritability of complex traits. Among these, rare and structural variants have been proposed as a type of genetic variation that is not captured by the existing designs and methodologies but plays an important role in the etiology and manifestation of complex traits, including SRD. Thus, in contrast to the CDCV hypothesis, the common disorder-rare variant (CDRV) hypothesis assumes that each case (or cluster of cases) is caused by a single rare variant characterized by a large effect size. These variants are individually rare in the general population, yet might cumulatively account for a sizeable proportion of cases. Interestingly, these variants can occur in different genes (or in different locations within the same gene) in different families/individuals. This hypothesis can explain the robustness of findings of the genetic underpinning for specific (and frequently severely affected) cases or families and the lack of generalizability of these findings to heterogeneous SRD samples. Given the methodological issues specific to studying rare variants (Kornilov & Grigorenko, 2016), the field has yet to produce a comprehensive examination of the role of this type of genetic variation in SRD, but the results from the first wave of whole-genome or whole-exome DNA sequencing and CNV association studies of SRD are encouraging. Two studies provide tentative support for the CDRV hypothesis as applied to SRD.

First, Gialluisi et al. (2016) used the data from the Illumina OmniExpress microarray platform (723,000 markers) that was specifically processed to obtain the CNV calls (i.e., establish the presence of the reduced or increased number of copies of genomic segments) and performed a CNV GWAS in a sample of 343 unrelated twinships/sibships that were part of the Colorado Learning Disabilities Research Centre cohort recruited through a proband with either SRD or attention-deficit/hyperactivity disorder. This study used a variety of analytical approaches and produced a complex pattern of tentative findings. First, the overall CNV burden (cumulative size of affected regions across the genome) did not correlate with the first principal component that was calculated for a set of reading and language measures. Second, no genome-wide significant signals were detected in the analysis that individually examined each marker. Several nominally significant associations were intriguing, however, including those for probes located in *CNTN4* (contactin 4; 3p26.3) and *CTNNA3* (catenin alpha 3; 10q21), genes that code for cell adhesion molecules hypothesized to play important roles in neuronal development, *CHRNA7* (cholinergic nicotinic receptor alpha 7; 15q13.3), and a marker located within close proximity of *ZNF737* (zinc finger protein 737, uncharacterized protein; 19p12). Although these results do not provide strong evidence for the role of rare variants in SRD from the viewpoint of the CDRV hypothesis, they point to a number of novel intriguing candidate genes and pathways, necessitating further large-scale studies of the role of rare variants in SRD.

In the second study, Einarsdottir et al. (2015) performed whole-exome sequencing of protein-coding regions of the DNA in a three-generation family with more than a third of family members affected with SRD (22 out of 62 individuals). The authors found several nonsynonymous (i.e., altering the protein composition through amino acid substitution) rare variants of potential interest, yet only one fully cosegregated with the disorder in this complex pedigree. This mutation affected the *CEP63* (centrosomal protein 63 kDa; 3q22.3) gene involved in cell division and highly expressed in the brain, possibly altering the protein structure or leading to the loss or reduction in protein complex formation with several CEP63-binding proteins. Moreover, subsequent validation attempts revealed that rs7619451, an SNP located within *CEP63*, was associated with white matter volume in the right hemisphere in healthy individuals, providing indirect support for the role of the gene in the development of the connectivity patterns implicated in complex cognition and reading. This study demonstrates the high utility of high-throughput next-generation DNA sequencing in identifying rare coding variants that might cause severe familial forms of SRD. Taken together, the two whole-exome and CNV studies previously mentioned represent the first step in the new direction of elucidating the genetic etiology of SRD, complementary to that taken by the GWAS.

We make two observations/commentaries to conclude this review of recent molecular genetic studies of SRD. First, CDCV and CDRV approaches are likely equally valid when applied to complex disorders such as SRD; yet, at the current stage of development of the field of human genetics and genomics of complex traits, little is known about their joint role in the etiology of common disorders in general and SRD in particular, and corresponding theoretical (as well as analytical) frameworks that integrate common and rare variant approaches are scarce. Second, only a few studies of the genetic contributions to SRD endophenotypes have been published. Although the patterns of results emerging from these interdisciplinary studies are intriguing, the majority of reported associations remain unexplored in independent cohorts or with respect to their role in the neurobiology of SRD. A set of such endophenotypes pertains to brain lateralization and handedness. Multiple studies inspired by the GGH have attempted to connect these endophenotypes to the etiology of SRD. Although the empirical results of these studies have been mixed, the intellectual power of the hypothesis keeps the inquiry open.

Little is still known about the specific roles of these genes (and corresponding variants) in the molecular and neurodevelopmental processes that contribute to the etiology and manifestation of SRD, despite the intriguing and rich pattern of suggestive associations revealed in the past couple of decades, in general, and in the last 5 years of GWAS, in particular, and the identification of a number of candidate SRD genomic regions, genes, and genetic variants. Attempts to replicate some of these novel results and candidates have not been fruitful (Carrion-Castillo et al., 2016), and caution should be exercised in interpreting this complex landscape of findings. The advancement of our knowledge about the etiology of SRD will require a careful integration of the statistical (e.g., genetic association) evidence and the theory-driven neurobiological approaches that attempt to illuminate the genetic architecture of SRD. The implications of this research should not be undervalued, and it will remain to be seen whether (and if yes, how) the progress that will be undoubtedly made in the next decade of research on molecular genetic bases of SRD can help further the development of its early diagnostics and personalized intervention.

REFERENCES

Brandler, W. M., Morris, A. P., Evans, D. M., Scerri, T. S., Kemp, J. P., Timpson, N. J., Paracchini, S. (2013). Common variants in left/right asymmetry genes and pathways are associated with relative hand skill. *PLoS Genetics, 9*(9), e1003751. doi:10.1371/journal.pgen.1003751

Brooks, A., Fulker, D. W., & DeFries, J. (1990). Reading performance and general cognitive ability: A multivariate genetic analysis of twin data. *Personality and Individual Differences, 11*, 141–146.

Byrne, B., Coventry, W. L., Olson, R. K., Samuelsson, S., Corley, R., Willcutt, E. G.,... Defries, J. C. (2009). Genetic and environmental influences on aspects of literacy and language in early childhood: Continuity and change from preschool to Grade 2. *Journal of Neurolinguistics, 22*, 219–236. doi:10.1016/j.jneuroling.2008.09.003

Cannon, T. D., & Keller, M. C. (2006). Endophenotypes in the genetic analyses of mental disorders. *Annual Review of Clinical Psychology, 2*, 267–290. doi:10.1146/annurev.clinpsy.2.022305.095232

Carrion-Castillo, A., Franke, B., & Fisher, S. E. (2013). Molecular genetics of dyslexia: An overview. *Dyslexia, 19*, 214–240. doi:10.1002/dys.1464

Carrion-Castillo, A., van Berge, E., Vino, A., van Zuijen, T. , de Jong, P. F., Francks, C., Fisher, S. E. (2016). Evaluation of results from genome-wide studies of language and reading in a novel independent dataset. *Genes, Brain, and Behavior, 15*(6), 531–541.

Eicher, J. D., Montgomery, A. M., Akshoomoff, N., Amaral, D. G., Bloss, C. S., Libiger, O.,…Gruen, J. R. (2016). Dyslexia and language impairment associated genetic markers influence cortical thickness and white matter in typically developing children. *Brain Imaging and Behavior, 10*(1), 272–282. doi:10.1007/s11682-015-9392-6

Eicher, J. D., Powers, N. R., Miller, L. L., Akshoomoff, N., Amaral, D. G., Bloss, C. S., … Gruen, J. R. (2013). Genome-wide association study of shared components of reading disability and language impairment. *Genes, Brain and Behavior, 12*(8), 792–801. doi:10.1111/gbb.12085

Einarsdottir, E., Svensson, I., Darki, F., Peyrard-Janvid, M., Lindvall, J., Ameur, A.,… Matsson, H. (2015). Mutation in CEP63 co-segregating with developmental dyslexia in a Swedish family. *Human Genetics, 134*(11-12), 1239–1248.

Friend, A., DeFries, J. C., & Olson, R. K. (2008). Parental education moderates genetic influences on reading disability. *Psychological Science, 19,* 1124–1130.

Geschwind, N., & Galaburda, A. M. (1985a). Cerebral lateralization. Biological mechanisms, associations, and pathology: I. A hypothesis and a program for research. *Archives of Neurology, 42*(5): 428–459.

Geschwind, N., & Galaburda, A. M. (1985b). Cerebral lateralization. Biological mechanisms, associations, and pathology: II. A hypothesis and a program for research. *Archives of Neurology, 42*(6), 521–552.

Geschwind, N., & Galaburda, A. M. (1985c). Cerebral lateralization. Biological mechanisms, associations, and pathology: III. A hypothesis and a program for research. *Archives of Neurology, 42*(7), 634–654.

Gialluisi, A., Guadalupe, T., Francks, C., & Fisher, S. E. (in press). Neuroimaging genetic analyses of novel candidate genes associated with reading and language. *Brain and Language.* doi:http://dx.doi.org/10.1016/j.bandl.2016.07.002

Gialluisi, A., Newbury, D. F., Wilcutt, E. G., Olson, R. K., DeFries, J. C., Brandler, W. M.,… Fisher, S. E. (2014). Genome-wide screening for DNA variants associated with reading and language traits. *Genes, Brain and Behavior, 13*(7), 686–701. doi:10.1111/gbb.12158

Gialluisi, A., Visconti, A., Willcutt, E. G., Smith, S. D., Pennington, B. F., Falchi, M.,… Fisher, S. E. (2016). Investigating the effects of copy number variants on reading and language performance. *Journal of Neurodevelopmental Disorders, 8,* 17. doi:10.1186/s11689-016-9147-8

Golding, J., & ALSPAC Study Team. (2004). The Avon Longitudinal Study of Parents and Children (ALSPAC) – study design and collaborative opportunities. *European Journal of Endocrinology, 151,* 119–123.

Grigorenko, E. L. (2004). Genetic bases of developmental dyslexia: A capsule review of heritability estimates. *Enfance, 3,* 273–287.

Grigorenko, E. L., Ngorosho, D., Jukes, M., & Bundy, D. (2006). Reading in able and disabled readers from around the world: Same or different? An illustration from a study of reading-related processes in a Swahili sample of siblings. *Journal of Reading Research, 29,* 104–123.

Jasińska, K. K., Molfese, P. J., Kornilov, S. A., Mencl, W. E., Frost, S. J., Lee, M.,…Landi, N. (2016). The BDNF Val66Met polymorphism influences reading ability and patterns of neural activation in children. *PLoS One, 11*(8), e0157449.

Jernigan, T. L., Brown, T. T., Hagler, D. J. Jr., Akshoomoff, N., Bartsch, H., Newman, E.,… Dale, A. M. (2016). Pediatric Imaging, Neurocognition, and Genetics (PING) Data Repository. *NeuroImage, 124(Pt B),* 1149–1154. Doi:10.1016/j.neuroimage.2015.04.057

Kornilov, S. A., & Grigorenko, E. L. (2016). Molecular genetics methods for developmental scientists. In D. Cicchetti (Ed), Development and Psychopathology, Volume Two: Developmental Neuroscience (3rd ed.) (pp. 378–415). NY: Wiley.

Landi, N., Frost, S. J., Mencl, W. E., Preston, J. L., Jacobsen, L. K., Lee, M.,…Grigorenko, E. L. (2013). The COMT val/met polymorphism is associated with reading-related skills and consistent patterns of functional neural activation. *Developmental Science, 16*(1), 13–23.

Luciano, M., Evans, D. M., Hansell, N. K., Medland, S. E., Montgomery, G. W., Martin, N. G.,...Bates, T. C. (2013). A genome-wide association study for reading and language abilities in two population cohorts. *Genes, Brain and Behavior, 12*(6), 645–652. doi:10.1111/gbb.12053

Manolio, T. A., Collins, F. S., Cox, N. J., Goldstein, D. B., Hindorff, L. A., Hunter, D. J.,... Visscher, P. M. (2009). Finding the missing heritability of complex diseases. *Nature, 461*(7265), 747–753.

Naples, A. J., Chang, J. T., Katz, L., & Grigorenko, E. L. (2009). Same or different? Insights into the etiology of phonological awareness and rapid naming. *Biological Psychology, 80,* 226–239.

Peterson, R. L., & Pennington, B. F. (2012). Developmental dyslexia. *The Lancet, 379,* 1997–2007.

Plomin, R., & Kovas, Y. (2005). Generalist genes and learning disabilities. *Psychological Bulletin, 131,* 592–617.

Roeske, D., Ludwig, K. U., Neuhoff, N., Becker, J., Bartling, J., Bruder, J.,...Schulte-Korne, G. (2011). First genome-wide association scan on neurophysiological endophenotypes points to trans-regulation effects on SLC2A3 in dyslexic children. *Molecular Psychiatry, 16,* 97–107. doi:10.1038/mp.2009.102

Schulte-Körne, G., Deimel, W., Muller, K., Gutenbrunner, C., & Remschmidt, H. (1996). Familial aggregation of spelling disability. *Journal of Child Psychology and Psychiatry and Allied Disciplines, 37,* 817–822.

Skeide, M. A., Kirsten, H., Kraft, I., Schaadt, G., Müller, B., Neef, N.,...Friederici, A. D. (2015). Genetic dyslexia risk variant is related to neural connectivity patterns underlying phonological awareness in children. *NeuroImage, 118,* 414–421.

Skiba, T., Landi, N., Wagner, R., & Grigorenko, E. L. (2011). In search of the perfect phenotype: An analysis of linkage and association studies of reading and reading-related processes. *Behavior Genetics, 41*(1), 6–30.

Taylor, J., Roehrig, A. D., Soden Hensler, B., Connor, C. M., & Schatschneider, C. (2010). Teacher quality moderates the genetic effects on early reading. *Science, 328,* 512–514.

Visscher, P. M., Brown, M. A., McCarthy, M. I., & Yang, J. (2012). Five years of GWAS discovery. *American Journal of Human Genetics, 90,* 7–24.

Wadsworth, S., Corley, R., Hewitt, J., & DeFries, J. (2001). Stability of genetic and environmental influences on reading performance at 7, 12, and 16 years of age in the Colorado Adoption Project. *Behavior Genetics, 31,* 353–359.

Wolff, P. H., & Melngailis, I. (1994). Family patterns of developmental dyslexia. *American Journal of Medical Genetics (Neuropsychiatric Genetics), 54,* 122–131.

CHAPTER 14

Intergenerational Transmission of Reading and Reading Brain Networks

Fumiko Hoeft and Roeland Hancock

SUMMARY

Intergenerational transmission (passing a phenotype from parents to offspring) of dyslexia and reading operates not only through heritable processes but also through genetic, epigenetic (changes in gene function in the absence of gene sequence changes), and environmental pathways. The integration of genetic, cognitive, and advanced neuroimaging methods within this context allows for clearer understanding about how these pathways interactively affect the development of the reading brain and mechanisms of intergenerational transmission, including sex-specific transmission and patterns at various stages of development. This chapter introduces the intergenerational neuroimaging approach, which is an approach that uses neuroimaging to study the relationship of cognitive and neural phenotypes between parents and their offspring. These studies are important in furthering our understanding of the Geschwind-Galaburda Hypothesis (GGH; Geschwind & Galaburda, 1985a, b, c), in which the impact of sex-specific effects at the prenatal stage are central. For example, studies are starting to reveal the impact of prenatal testosterone on brain development, epigenetic effects of estrogen on dyslexia susceptibility genes, and sex-specific transmission patterns in reading-related brain circuits in prereaders. This research can ultimately affect early identification and intervention in dyslexia by understanding which parental factors affect dyslexia and their associated mechanisms with greater precision. Such research will also allow identification of modifiable targets, such as prenatal nutrition and maternal smoking, for early preventive interventions.

INTRODUCTION

Dyslexia is a complex trait that develops from multiple interacting genetic, neural, and cognitive factors and environmental interactions with these factors (Pennington, 2006). Studies of intergenerational transmission in dyslexia may unravel these complex interactions and help understand the biological mechanisms of dyslexia. *Intergenerational transmission* refers to a phenotype passing from parents to offspring through mechanisms such as heritable genetic influences and post- and prenatal environmental effects.

Prenatal environmental effects are central to the GGH (Geschwind & Galaburda, 1985a, b, c), a theory of hormonal effects on brain development and lateralization. The GGH was motivated in part by the observation that the sex-specific risk of developmental disorders associated with left-hemisphere brain function, such as dyslexia, is not consistent with sex chromosome linkage, but is generally higher in males than in females. These observations suggested a prenatal impact on brain development in dyslexia mediated by sex differences in in utero testosterone. Although a direct relationship between prenatal testosterone and dyslexia has not been supported (Boets, De Smedt, Wouters, Lemay, & Ghesquière, 2007), evidence shows that testosterone mediates brain plasticity in response to induced ectopias that are similar to those seen in dyslexia (Rosen et al., 2007).

Ectopias, which are small aggregations of neurons in the superficial layer of the cortex, were primarily identified in perisylvian regions in postmortem case studies of brains from individuals with severe reading difficulties (Galaburda, Sherman, Rosen, Aboitiz, & Geschwind, 1985). Ectopias are presumably due to genetic or environmental factors during early fetal brain development because neuronal migration is thought to occur as early as the second trimester. The effect of testosterone on brain development may also occur later during the perinatal period (see Chapter 1).

Intergenerational transmission research may clarify the role of sex-specific epigenetic and prenatal environmental effects (e.g., testosterone) and how these factors may interact with genetic factors. We emphasize neuroimaging approaches in the examination of intergenerational transmission of the human brain circuitry, which we refer to as *intergenerational neuroimaging* (Ho, Sanders, Gotlieb, & Hoeft, 2016).

INTERGENERATIONAL TRANSMISSION OF READING

Intergenerational, multifactorial models of dyslexia (van Bergen, van der Leij, & de Jong, 2014) are supported by studies showing a relationship between parental reading history and their offsprings' liability for developing dyslexia. In a prospective longitudinal study, parents with dyslexia whose children developed dyslexia had poorer reading fluency, not explained by environmental measures, compared with those whose children did not develop dyslexia (van Bergen et al., 2011; see also Torppa, Eklund, van Bergen, & Lyytinen, 2011), supporting the idea that the parents' reading history confers liability to their children for developing dyslexia.

Neuroimaging studies that compare prereaders with and without a familial risk for dyslexia complement these behavioral findings (see Chapter 15). Brain differences between high- and low-risk prereaders are most consistently found in left temporo-parietal regions, which are typically implicated in phonological processing and grapheme-phoneme conversion, followed by the left occipito-temporal region, which is important for orthographic processing (Vandermosten, Hoeft, & Norton, 2016). There is also evidence for anomalies in the auditory domain (Clark et al., 2014; Raschle, Stering, Meissner, & Gaab, 2014) and in brain connectivity (Hosseini et al., 2013; Saygin et al., 2013; Vandermosten et al., 2015) in high-risk prereaders.

These studies show intergenerational transmission of reading and related brain phenotypes. These designs, however, do not differentiate sources of intergenerational transmission (i.e., genetics, prenatal environment, postnatal environment). There are designs that can differentiate these factors, which are described below and in a recent article by our group (Ho et al., 2016).

INTERGENERATIONAL TRANSMISSION IN THE BRAIN

Embryonic cortical development is controlled by systematic differences in proteins known as morphogens and transcription factors across the neural plate that forms the basis for the nervous system (see Chapter 3). This results in brain regions with shared genetic influences, which lead to the orderly phenotypic organization of the brain (Chen et al., 2013). Each of these genetically organized regions may have distinct points of intergenerational influence that can contribute to reading-related and other cognitive phenotypes. For example, inconsistent findings of abnormalities in either the temporo-parietal or occipito-temporal regions in high-risk prereaders (see Vandermosten et al., 2016), which correspond to different patterns of gene expression, may reflect the outcomes of distinct genetic variants or environmental effects during brain development.

Twin studies reported moderate to high heritability for structural brain measures (see Blokland, de Zubicaray, McMahon, & Wright, 2012, for a review and meta-analysis). Further research suggests that cortical thickness and surface area, which are highly heritable, are controlled by largely independent genetic mechanisms (Chen et al., 2013), and may reflect distinct developmental time courses of these two measures. Meta-analysis and a relatively large neuroimaging sample of 476 healthy adults also support high heritability of white matter integrity (Kochunov et al., 2015). Functional brain activation is also moderately heritable and possibly influenced by genetic factors distinct from those controlling brain structure (Glahn et al., 2010).

Twin studies can produce systematically biased heritability estimates. Notably, twin studies assume that twins share common prenatal environmental influences, but there is intertwin variability in the distribution of blood and nutrients in utero for both monozygotic (MZ) and dizygotic (DZ) twins (Marceau et al., 2016). In a unique study taking advantage of natural variability in birth weight between twins, in utero factors had long-term effects on brain maturation (Alexander-Bloch, Raznahan, Bullmore, & Giedd, 2013). Birth weight is a proxy for some prenatal environmental factors, including testosterone (Carlsen, Jacobsen, & Romundstad, 2006), and studies using direct measures of prenatal testosterone assayed from amniotic fluid also found that testosterone influences brain structure (Chura et al., 2010; Lombardo et al., 2012). These studies highlight the significant role of prenatal environmental factors on brain development and provide support for the core of the GGH.

Sources of Intergenerational Transmission of Reading and Related Brain Circuits

Postnatal influences on reading are difficult to separate from other sources of intergenerational transmission, although the available evidence suggests that parent–offspring similarities in reading are not primarily due to postnatal effects. Wadsworth, Corley, Hewitt, Plomin, and DeFries (2002) found a significant association between reading ability in biologically related parents and their children but not in adoptive parents and their adopted children. Swagerman et al. (2015) found no evidence for parent–offspring cultural transmission of reading ability. These studies together suggest minimal postnatal intergenerational transmission of reading.

Genetics Dyslexia is a complex trait with heritability estimates of .41–.74 (Grigorenko, 2004). A number of dyslexia susceptibility genes have been identified and play a putative role in neuronal migration, axon guidance, and dendritic morphology (*KIAA0319:* Paracchini et al., 2006; *DCDC2:* Burbridge et al., 2008; Meng et al., 2005; *DYX1C1:* Wang et al., 2006; *ROBO1:* Hannula-Jouppi et al., 2005; see Chapter 3). These findings complement reports of ectopias in dyslexia, which are likely the result of disrupted neuronal migration (Galaburda et al., 1985).

Imaging genetics studies, which link genetic variation, candidate neural endophenotypes, and behavior, have primarily identified associations between risk single nucleotide polymorphisms and the structure and function of the left temporo-parietal and occipito-temporal regions (Darki, Peyrard-Janvid, Matsson, Kere, & Klingberg, 2012; Meda et al., 2008; Meng et al., 2005). Many of the studies used a region-of-interest approach to maximize power and have focused on well-established genetic loci, however, and may have missed associations with other regions and other genetic loci potentially important for dyslexia (e.g., *SLC2A3:* Roeske et al., 2011). Larger replication studies are needed to establish evidence that the effect of genetic risk variants on behavioral phenotypes is mediated by neural measures.

In addition to studies that examined heritability across the brain, a few studies focused on heritability for reading-specific brain function, with discrepant results. Polk, Park, Smith, and Park (2007) found that adult MZ twins showed greater similarity than DZ twins for gray matter volume within both the visual word form area (VWFA) and fusiform face areas (FFA), which dissociated from similarities for functional activation in response to words and faces, with only face-related activation showing significantly greater MZ twin similarity. This result suggests that orthographic processing is not strongly heritable at a neurofunctional level, although it may be anatomically constrained by heritable factors. Yet, Pinel et al. (2015) reported nearly the opposite finding—functional activation in the left VWFA, but not FFA, was more similar in MZ than DZ twins. Both studies employed small samples of adult readers, so more research is needed to replicate these findings and resolve questions over the heritability of orthographic processing.

GxE Interaction and Epigenetics *GxE interaction* typically refers to a phenomenon that is widely documented in various neurodevelopmental disorders in which sensitivity to environmental risk factors interacts with one's genotype (Caspi & Moffitt, 2006). As in other complex traits, known genetic risk variants account for only a small fraction of the heritability in reading suggested by twin studies. Several issues may underlie this missing heritability, including epistatic (gene–gene interaction) effects, limited power to identify polygenic inheritance, and substantial GxE interaction effects in reading and reading-related brain networks. These pose a considerable challenge to genetic studies of reading when pre- and postnatal and biological environmental effects are poorly characterized.

Most behavioral genetics studies in dyslexia show greater genetic influence in more advantaged environments (measured by parental education) compared with less advantaged environments (Friend, DeFries, & Olson, 2008). There is also support for the diathesis-stress (Rende & Plomin, 1992) and resilience models (Pennington et al., 2009) of GxE interactions, however, which posit that genetic influence is greater in less optimal environments. Mascheretti et al. (2013) found that the effects of the *DYX1C1* gene on reading and phonological memory in children with dyslexia and their siblings were dependent on environmental risk factors such as birth weight, socioeconomic status (SES), and maternal smoking during pregnancy. Although several neuroimaging studies examined the effect of environment, often using SES as a surrogate for environment, on reading-related brain functional and structural patterns (Jednoróg et al., 2012; Noble et al., 2015; Piccolo, Merz, He, Sowell, & Noble, 2016; Raizada & Kishiyama, 2010), only a couple of studies examined GxE interactions, often using a dyslexia diagnosis and family history as proxies for genetic effects. One study found activation patterns in dyslexia to be similar, regardless of SES, suggesting no GxE interaction (Monzalvo, Fluss, Billard, Dehaene, & Dehaene-Lambertz, 2012), whereas another study suggested GxE interaction; high familial risk prereaders showed weaker association between home literacy environment and brain activation in phonological and orthographic processing-related regions, which was interpreted as a weaker environmental effect with greater genetic liability (Powers, Wang, Beach, Sideridis, & Gaab, 2016).

Environmental factors can also trigger epigenetic modifications (i.e., heritable molecular changes that can affect phenotypes through changes in gene expression without changing the underlying genome). Epigenetic modifications have a significant role in intergenerational transmission through inherited modifications and environmentally induced modifications, such as those that can occur prenatally. The influence of epigenetic changes on neurodevelopmental disorders is an important topic that is just starting to be addressed in dyslexia (Smith, 2011). Mutations associated with reading have been linked to changes in transcription factor binding and gene expression (e.g., *DYX1C1:* Tapia-Páez, Tammimies, Massinen, Roy, & Kere, 2008; *KIAA0319:* Paracchini et al., 2006), suggesting that gene expression levels, rather than changes in gene structure, may account for some variance in reading. Only one epigenetic study relating to reading has been published, which found that *DYX1C1* expression is regulated by estrogen, possibly through epigenetic methylation (Paracchini et al., 2006). Epigenetic effects may also vary according to the sex of the offspring and whether the inherited gene is received from the mother or from the father—a parent-of-origin effect (Mott et al., 2014). Most complex traits may show some form of parent-of-origin effects (Lawson, Cheverud, & Wolf, 2013). Only one study so far has found parent-of-origin effects in reading circuitry, showing that maternal, but not paternal, reading history is negatively correlated with prereading measures and temporo-parietal surface area (Black et al., 2012). Together with animal work showing that genetic defects in histone modification could be reversed using therapeutic interventions (see Smith, 2011 for a review), research on epigenetics holds great promise in understanding the influence of environment on genotypes, understanding intergenerational transmission patterns, and as an avenue for future development of interventions.

CONCLUSION

We have discussed past research on how traits such as reading and dyslexia are passed on from parents to offspring (i.e., intergenerational transmission) and how future research on intergenerational transmission can help us understand factors that affect these traits. Studies established the high heritability of dyslexia and reading. Studies are also starting to establish how intergenerational transmission operates through genetic, epigenetic and environmental pathways. The integration of genetic, cognitive and advanced neuroimaging methods allows us to elucidate intergenerational transmission of brain circuitry involved in reading and dyslexia. These advances promise to shed insight into possible mechanisms of transmission.

Future research on mechanisms underlying dyslexia is necessary, particularly in areas of epigenetics and sex-specific transmission patterns, which may shed further light on the GGH. Other future research areas include addressing missing heritability issues, such as examination of the role for genetic factors other than common variants (e.g., de novo mutations) or more complex GxE and epistatic effects. Finally, there is now the potential for natural cross-fostering designs utilizing different types of in vitro fertilization families to disentangle inherited factors from pre- and postnatal influences (Ho et al., 2016). This design is particularly valuable in isolating prenatal influence in humans, which is important in understanding the GGH. We ultimately believe that this line of research can affect early identification and intervention of dyslexia. By examining intergenerational transmission patterns, we may be able to identify the probability with which these parental factors may confer risk to their children with greater precision. It will also allow identification of modifiable targets, such as prenatal nutrition and maternal smoking, for early preventive interventions.

REFERENCES

Alexander-Bloch, A., Raznahan, A., Bullmore, E., & Giedd, J. (2013). The convergence of maturational change and structural covariance in human cortical networks. *Journal of Neuroscience: The Official Journal of the Society for Neuroscience, 33*(7), 2889–2899.

Black, J. M., Tanaka, H., Stanley, L., Nagamine, M., Zakerani, N., Thurston, A.,…Hoeft, F. (2012). Maternal history of reading difficulty is associated with reduced language-related gray matter in beginning readers. *NeuroImage, 59*(3), 3021–3032.

Blokland, G. A., de Zubicaray, G. I., McMahon, K. L., & Wright, M. J. (2012). Genetic and environmental influences on neuroimaging phenotypes: A meta-analytical perspective on twin imaging studies. *Twin Research and Human Genetics, 15*(3), 351–371.

Boets, B., De Smedt, B., Wouters, J., Lemay, K., & Ghesquière, P. (2007). No relation between 2D:4D fetal testosterone marker and dyslexia. *NeuroReport, 18*(14), 7–11.

Burbridge, T. J., Wang, Y., Volz, A. J., Peschansky, V. J., Lisann, L., Galaburda, A. M.,… Rosen, G. D. (2008). Postnatal analysis of the effect of embryonic knockdown and overexpression of candidate dyslexia susceptibility gene homolog Dcdc2 in the rat. *Neuroscience, 152*(3), 723–733.

Carlsen, S. M., Jacobsen, G., & Romundstad, P. (2006). Maternal testosterone levels during pregnancy are associated with offspring size at birth. *European Journal of Endocrinology/ European Federation of Endocrine Societies, 155*(2), 365–370.

Caspi, A., & Moffitt, T. E. (2006). Gene-environment interactions in psychiatry: Joining forces with neuroscience. *Nature Reviews Neuroscience, 7*(7), 583–590.

Chen, C.-H., Fiecas, M., Gutierrez, E. D., Panizzon, M. S., Eyler, L. T., Vuoksimaa, E.,… Jernigan, T. L. (2013). Genetic topography of brain morphology. *Proceedings of the National Academy of Sciences, 110*(42), 17089–17094.

Chura, L. R., Lombardo, M. V., Ashwin, E., Auyeung, B., Chakrabarti, B., Bullmore, E. T., & Baron-Cohen, S. (2010). Organizational effects of fetal testosterone on human corpus callosum size and asymmetry. *Psychoneuroendocrinology, 35*(1), 122–132.

Clark, K. A., Helland, T., Specht, K., Narr, K. L., Manis, F. R., Toga, A. W., & Hugdahl, K. (2014). Neuroanatomical precursors of dyslexia identified from pre-reading through to age 11. *Brain, 137*(12), 3136–3141.

Darki, F., Peyrard-Janvid, M., Matsson, H., Kere, J., & Klingberg, T. (2012). Three dyslexia susceptibility genes, DYX1C1, DCDC2, and KIAA0319, affect temporo-parietal white matter structure. *Biological Psychiatry, 72*(8), 671–676.

Friend, A., DeFries, J. C., & Olson, R. K. (2008). Parental education moderates genetic influences on reading disability. *Psychological Science, 19*(11), 1124–1130.

Galaburda, A. M., Sherman, G. F., Rosen, G. D., Aboitiz, F., & Geschwind, N. (1985). Developmental dyslexia: Four consecutive patients with cortical anomalies. *Annals of Neurology, 18*(2), 222–233.

Geschwind, N., & Galaburda, A. M. (1985a). Cerebral lateralization. Biological mechanisms, associations, and pathology: I. A hypothesis and a program for research. *Archives of Neurology*, 42(5), 428–459.

Geschwind, N., & Galaburda, A. M. (1985b). Cerebral lateralization. Biological mechanisms, associations, and pathology: II. A hypothesis and a program for research. *Archives of Neurology, 42*(6), 521–552.

Geschwind, N., & Galaburda, A. M. (1985c). Cerebral lateralization. Biological mechanisms, associations, and pathology: III. A hypothesis and a program for research. *Archives of Neurology, 42*(7), 634–654.

Glahn, D. C., Winkler, A. M., Kochunov, P., Almasy, L., Duggirala, R., Carless, M. A.,…Blangero, J. (2010). Genetic control over the resting brain. *Proceedings of the National Academy of Sciences of the United States of America, 107*(3), 1223–1228.

Grigorenko, E. L. (2004). Genetic bases of developmental dyslexia: A capsule review of heritability estimates. *Enfance, 56*, 273–288.

Hannula-Jouppi, K., Kaminen-Ahola, N., Taipale, M., Eklund, R., Nopola-Hemmi, J., Kääriäinen, H., & Kere, J. (2005). The axon guidance receptor gene ROBO1 is a candidate gene for developmental dyslexia. *PLoS Genetics, 1*(4), e50.

Ho, T. C., Sanders, S., Gotlieb, I. H., & Hoeft, F. (2016). Intergenerational neuroimaging of human brain circuitry. *Trends in Neuroscience, 39*(10), 644–648.

Hosseini, S. M. H., Black, J. M., Soriano, T., Bugescu, N., Martinez, R., Raman, M. M.,…Hoeft, F. (2013). Topological properties of large-scale structural brain networks in children with familial risk for reading difficulties. *NeuroImage, 71*, 260–274.

Jednoróg, K., Altarelli, I., Monzalvo, K., Fluss, J., Dubois, J., Billard, C.,…Ramus, F. (2012). The influence of socioeconomic status on children's brain structure. *PloS One, 7*(8), e42486.

Kochunov, P., Jahanshad, N., Marcus, D., Winkler, A., Sprooten, E., Nichols, T. E.,…Van Essen, D. C. (2015). Heritability of fractional anisotropy in human white matter: A comparison of Human Connectome Project and ENIGMA-DTI data. *NeuroImage, 111*(C), 300–311.

Lawson, H. A., Cheverud, J. M., & Wolf, J. B. (2013). Genomic imprinting and parent-of-origin effects on complex traits. *Nature Reviews Genetics, 14*(9), 609–617.

Lombardo, M. V., Ashwin, E., Auyeung, B., Chakrabarti, B., Taylor, K., Hackett, G.,…Baron-Cohen, S. (2012). Fetal testosterone influences sexually dimorphic gray matter in the human brain. *Journal of Neuroscience: The Official Journal of the Society for Neuroscience, 32*(2), 674–680.

Marceau, K., McMaster, M. T., Smith, T. F., Daams, J. G., Beijsterveldt, C. E., Boomsma, D. I., & Knopik, V. S. (2016). The prenatal environment in twin studies: a review on chorionicity. *Behavior Genetics, 46*(3), 286–303.

Mascheretti, S., Bureau, A., Battaglia, M., Simone, D., Quadrelli, E., Croteau, J.,…Marino, C. (2013). An assessment of gene-by-environment interactions in developmental dyslexia-related phenotypes. *Genes, Brain, and Behavior, 12*(1), 47–55.

Meda, S. A., Gelernter, J., Gruen, J. R., Calhoun, V. D., Meng, H., Cope, N. A., & Pearlson, G. D. (2008). Polymorphism of DCDC2 reveals differences in cortical morphology of healthy individuals: A preliminary voxel based morphometry study. *Brain Imaging and Behavior, 2*(1), 21–26.

Meng, H., Smith, S. D., Hager, K., Held, M., Liu, J., Olson, R. K.,…Gruen, J. R. (2005). DCDC2 is associated with reading disability and modulates neuronal development in the brain. *Proceedings of the National Academy of Sciences of the United States of America, 102*(47), 17053–17058.

Monzalvo, K., Fluss, J., Billard, C., Dehaene, S., & Dehaene-Lambertz, G. (2012). Cortical networks for vision and language in dyslexic and normal children of variable socioeconomic status. *NeuroImage, 61*(1), 258–274.

Mott, R., Yuan, W., Kaisaki, P., Gan, X., Cleak, J., Edwards, A.,…Flint, J. (2014). The architecture of parent-of-origin effects in mice. *Cell, 156*(1), 332–342.

Noble, K. G., Houston, S. M., Brito, N. H., Bartsch, H., Kan, E., Kuperman, J. M.,…Libiger, O. (2015). Family income, parental education and brain structure in children and adolescents. *Nature Neuroscience, 18*(5), 773–778.

Paracchini, S., Thomas, A., Castro, S., Lai, C., Paramasivam, M., Wang, Y.,…Monaco, A. P. (2006). The chromosome 6p22 haplotype associated with dyslexia reduces the expression of KIAA0319, a novel gene involved in neuronal migration. *Human Molecular Genetics, 15*(10), 1659–1666.

Pennington, B. F. (2006). From single to multiple deficit models of developmental disorders. *Cognition, 101*(2), 385–413.

Pennington, B. F., McGrath, L. M., Rosenberg, J., Barnard, H., Smith, S. D., Willcutt, E. G.,...Olson, R. K. (2009). Gene X environment interactions in reading disability and attention-deficit/hyperactivity disorder. *Developmental Psychology, 45*(1), 77-89.

Piccolo, L. R., Merz, E. C., He, X., Sowell, E. R., & Noble, K. G. (2016). Age-related differences in cortical thickness vary by socioeconomic status. *PloS One, 11*(9), e0162511.

Pinel, P., Lalanne, C., Bourgeron, T., Fauchereau, F., Poupon, C., Artiges, E.,...Dehaene, S. (2015). Genetic and environmental influences on the visual word form and fusiform face areas. *Cerebral Cortex, 25*(9), 2478–2493.

Polk, T. A., Park, J., Smith, M. R., & Park, D. C. (2007). Nature versus nurture in ventral visual cortex: A functional magnetic resonance imaging study of twins. *Journal of Neuroscience, 27*(51), 13921–13925.

Powers, S. J., Wang, Y., Beach, S. D., Sideridis, G. D., & Gaab, N. (2016). Examining the relationship between home literacy environment and neural correlates of phonological processing in beginning readers with and without a familial risk for dyslexia: An fMRI study. *Annals of Dyslexia,66*(3), 1–24.

Raizada, R. D. S., & Kishiyama, M. M. (2010). Effects of socioeconomic status on brain development and how cognitive neuroscience may contribute to levelling the playing field. *Frontiers in Human Neuroscience,* 4(3).

Raschle, N. M., Stering, P. L., Meissner, S. N., & Gaab, N. (2014). Altered neuronal response during rapid auditory processing and its relation to phonological processing in prereading children at familial risk for dyslexia. *Cerebral Cortex,* 24(9), 2489–2501.

Rende, R., & Plomin, R. (1992). Diathesis-stress models of psychopathology: A quantitative genetic perspective. *Applied and Preventive Psychology,* 1(4), 177–182.

Roeske, D., Ludwig, K. U., Neuhoff, N., Becker, J., Bartling, J., Bruder, J.,...Hoffmann, P. (2011). First genome-wide association scan on neurophysiological endophenotypes points to trans-regulation effects on SLC2A3 in dyslexic children. *Molecular Psychiatry, 16*(1), 97–107.

Rosen, G. D., Bai, J., Wang, Y., Fiondella, C. G., Threlkeld, S. W., LoTurco, J. J., & Galaburda, A. M. (2007). Disruption of neuronal migration by RNAi of Dyx1c1 results in neocortical and hippocampal malformations. *Cerebral Cortex, 17*(11), 2562–2572.

Saygin, Z. M., Norton, E. S., Osher, D. E., Beach, S. D., Cyr, A. B., Ozernov-Palchik, O.,...Gabrieli, J. D. (2013). Tracking the roots of reading ability: White matter volume and integrity correlate with phonological awareness in prereading and early-reading kindergarten children. *Journal of Neuroscience, 33*(33), 13251–13258.

Smith, S. D. (2011). Approach to epigenetic analysis in language disorders. *Journal of Neurodevelopmental Disorders,* 3(4), 356–364.

Swagerman, S. C., van Bergen, E., Dolan, C., de Geus, E. J., Koenis, M. M., Pol, H. E. H., & Boomsma, D. I. (2015). Genetic transmission of reading ability. *Brain and Language 97,* 32–39.

Tammimies, K., Tapia-Páez, I., Rüegg, J., Rosin, G., Kere, J., Gustafsson, J.-Å., & Nalvarte, I. (2012). The rs3743205 SNP is important for the regulation of the dyslexia candidate gene DYX1C1 by estrogen receptor β and DNA methylation. *Molecular Endocrinology, 26*(4), 619–629.

Tapia-Páez, I., Tammimies, K., Massinen, S., Roy, A. L., & Kere, J. (2008). The complex of TFII-I, PARP1, and SFPQ proteins regulates the DYX1C1 gene implicated in neuronal migration and dyslexia. *The FASEB Journal, 22*(8), 3001–3009.

Torppa, M., Eklund, K., van Bergen, E., & Lyytinen, H. (2011). Parental literacy predicts children's literacy: A longitudinal family-risk study. *Dyslexia, 17*(4), 339–355.

van Bergen, E., de Jong, P. F., Regtvoort, A., Oort, F., van Otterloo, S., & van der Leij, A. (2011). Dutch children at family risk of dyslexia: Precursors, reading development, and parental effects. *Dyslexia, 17*(1), 2–18.

van Bergen, E., van der Leij, A., & de Jong, P. F. (2014). The intergenerational multiple deficit model and the case of dyslexia. *Frontiers in Human Neuroscience, 8,* 346.

Vandermosten, M., Hoeft, F., & Norton, E. S. (2016). Integrating MRI brain imaging studies of pre-reading children with current theories of developmental dyslexia: A review and quantitative meta-analysis. *Current Opinion in Behavioral Sciences, 10,* 155–161.

Vandermosten, M., Vanderauwera, J., Theys, C., De Vos, A., Vanvooren, S., Sunaert, S.,… Ghesquière, P. (2015). A DTI tractography study in pre-readers at risk for dyslexia. *Developmental Cognitive Neuroscience, 14,* 8–15.

Wadsworth, S. J., Corley, R. P., Hewitt, J. K., Plomin, R., & DeFries, J. C. (2002). Parent–offspring resemblance for reading performance at 7, 12 and 16 years of age in the Colorado Adoption Project. *Journal of Child Psychology and Psychiatry, 43*(6), 769–774.

Wang, Y., Paramasivam, M., Thomas, A., Bai, J., Kaminen-Ahola, N., Kere, J.,… & Loturco, J. J. (2006). DYX1C1 functions in neuronal migration in developing neocortex. *Neuroscience, 143*(2), 515-522.

CHAPTER 15

Early Atypical Brain Development in Developmental Dyslexia

Nadine Gaab, Xi Yu, and Ola Ozernov-Palchik

SUMMARY

Significant technological advances since the publication of the Geschwind-Galaburda Hypothesis (GGH; Geschwind & Galaburda, 1985a, b, c) have made it possible to examine brain development in the living human brain. Results across numerous studies suggest atypical early structural and functional brain development in children with a familial risk or a diagnosis of developmental dyslexia, although the sensitivity and specificity of the observed brain changes is still under investigation. Atypical brain development is seen in both hemispheres of the brain, although the strongest effects are observed in regions within the left hemisphere. In addition, research suggests that there may be a network in the right hemisphere in some at-risk children that facilitates the development of typical reading skills. Overall, research results since the mid-1980s partially support the GGH, but they show a more complex picture of how children's brains develop and its hemispheric specialization. The GGH has triggered a vast amount of work on early brain development in learning disorders and has led to a developmental neurobiological approach to studying dyslexia. This approach combines behavioral and brain research and will continue to play a crucial role in the development of effective strategies for early identification and intervention. Most important, it may help to reduce the devastating clinical, psychological, and social consequences of developmental dyslexia.

INTRODUCTION

No viable methods were available for noninvasive in vivo measuring of brain activation and brain structure with high spatial precision when Norman Geschwind and Albert Galaburda (1985a, b, c) published their seminal theory on the neurobiological mechanisms of cerebral lateralization and the role of these mechanisms in developmental disorders such as dyslexia. The advent of magnetic resonance imaging (MRI) technology allowed for safe and comprehensive investigation of typical and atypical neuroanatomy across perceptual and cognitive domains, including reading (Raschle et al. 2009; Raschle et al., 2012). Extensive discoveries have been made on the neurobiological basis of developmental dyslexia since the mid-1990s (Gabrieli, 2009; Norton, Beach, & Gabrieli, 2015), and Geschwind and Galaburda's original reports from postmortem studies of individuals with dyslexia were affirmed: Differences in brain anatomy can be observed in individuals with developmental dyslexia across the developmental timeline.

The vast majority of these studies, however, were conducted in school-age children and adults diagnosed with developmental dyslexia. The limitation of this approach was alluded to in the GGH: "Although genetic factors are important, we will lay stress on several factors that, in the course of development, both prenatal and postnatal, modify the direction and extent of these structural differences" (Geschwind & Galaburda, 1985a, p. 428). The main limitation of neuroimaging studies in children who had already been instructed in reading (e.g., second graders and older) is that they represented both the innate neurobiological deficits in the brains of readers with dyslexia and the consequences of environmentally driven neuroplasticity (i.e., the conflated effects of struggling to learn to read) (Goswami, 2015). Several etiological mechanisms (e.g., genetics, neurobiology, perceptual and cognitive deficits, and environment) can contribute to reading failure (Ozernov-Palchik, Yu, Wang, & Gaab, 2016). Examining the contributing factors across all stages of the developmental trajectory of learning to read, from birth to adulthood, is important to understanding the etiology of developmental dyslexia.

This chapter briefly reviews the neurobiological factors of developmental dyslexia and summarizes research findings from MRI studies of pre-reading children at risk for dyslexia. The studies are connected with two GGH predictions—atypical brain development and atypical lateralization in developmental dyslexia.

STRUCTURAL AND FUNCTIONAL BRAIN ATYPICALITIES IN DYSLEXIA

Developmental dyslexia has been associated with atypicalities in brain regions important for reading. MRI studies in children and adults with developmental dyslexia commonly demonstrate reduced gray matter volume and cortical thickness, as well as hypoactivation in bilateral temporoparietal and inferior frontal and left occipito-temporal networks (Ozernov-Palchik et al., 2016). Furthermore, atypical functional and structural connectivity among these regions has been demonstrated (e.g., Boets et al., 2013; Horwitz, Rumsey, & Donohue, 1998; Morken, Helland, Hugdahl, & Specht, 2017; Rimrodt, Peterson, Denckla, Kaufmann, & Cutting, 2010; Stanberry et al., 2006; van der Mark et al., 2011). Brains of individuals with developmental dyslexia have further been characterized by reduced asymmetry in posterior brain regions important for language processing, such as the planum temporale or other areas surrounding the sylvian fissure (e.g., Altarelli et al., 2014; Galaburda, Menard, & Rosen, 1994).

Brain Atypicalities in Preschoolers at Risk for Dyslexia

A key question in research on the neural basis of dyslexia concerns which brain characteristics of developmental dyslexia may be related to the cause of the reading difficulty versus the result of reduced reading experience due to struggles in learning to read. In order to investigate this question, we conducted a series of longitudinal and cross-sectional studies in children who are prereaders and infants with hereditary risk for developmental dyslexia. Studies estimate a 50% risk for developing dyslexia in children with a first-degree relative with the disorder (Ziegler et al., 2005); therefore, this group provides a unique opportunity to examine early, genetically and prenatally determined atypicalities in brain development, as well as disentangle brain characteristics in developmental dyslexia that are postnatally determined.

Structure and Function

We demonstrated hypoactivation in bilateral occipito-temporal and left temporo-parietal regions during phonological processing in an investigation of children who are prereaders with a family history for developmental dyslexia (FHD+) and without (FHD–) (Raschle, Stering, Meissner, & Gaab, 2013; Raschle, Zuk, & Gaab, 2012). In addition, reduced gray matter volume indices in bilateral occipito-temporal and temporo-parietal regions were observed in FHD+ compared to FHD– preschoolers, and gray matter indices in these left-hemispheric regions were positively correlated with prereading skills (Raschle, Chang, & Gaab, 2011). We also observed significantly decreased similarity (compared to typical developing children) in the sulcal basin area in left temporo-parietal and occipito-temporal regions in children with developmental dyslexia as well as younger FHD+ preschoolers (Im, Raschle, Smith, Grant, & Gaab, 2015). The degree of similarity of sulcal patterns correlated positively with reading performance, supporting the idea of atypical early brain development in developmental dyslexia starting in utero because sulcal pattern is primarily determined prenatally (Chi, Dooling, & Gilles, 1977; Kostovic & Vasung, 2009).

Structural Connectivity

We demonstrated atypical development of white matter pathways, specifically of the arcuate fasciculus (AF) and the inferior and superior longitudinal fasciculi (ILF and SLF, respectively), from the prereading to the fluent reading stage in FHD+ compared to FHD– children (Wang et al., 2017). An automated fiber quantification (AFQ) method (Yeatman, Dougherty, Myall, Wandell, & Feldman, 2012) was employed in this DTI study to quantify the fractional anisotropy (FA) of multiple nodes along the three tracts of interest important for reading. The FA development rate of the left AF significantly differed between children who subsequently developed into good or poor readers, and was positively correlated with gains in reading performance. The rate of FA development in the SLF together with familial risk and prereading performance further predicted reading fluency subsequently once the children entered elementary school.

In addition, hemispheric lateralization of the three tracts was examined. FHD+ children showed right lateralization in AF at the prereading and the beginning reading stages, whereas FHD– children showed left lateralization in AF, with the group differences being most prominent at the prereading stage. This result supports previous studies that showed reduced leftward asymmetry in posterior white matter regions and the AF in children with reading deficits (Niogi & McCandliss, 2006; Vandermosten et al., 2015). We conducted another study (Project READ) of prereaders in collaboration with researchers at the Massachusetts Institute of Technology and demonstrated that children who had lower phonological awareness scores had significantly reduced volume and fractional anisotropy of the left AF, but not in other tracts (Saygin et al., 2013). These findings indicate atypical white matter development associated with hereditary and behavioral risk for dyslexia. It further supports reduced left-lateralization in white matter tracts important for reading development in individuals with dyslexia.

Studies from other laboratories have corroborated our findings of atypical white matter development in developmental dyslexia using various methods (Black et al., 2012; Clark et. al., 2014; Myers et. al., 2014; van Herten et al., 2008; Vandermosten et al., 2015). For example, in one study in Dutch speakers, FHD+ kindergartners demonstrated significantly reduced FA in the left inferior fronto-occipital fasciculus (IFOF), and increased FA in IFOF bilaterally was associated with phonological awareness (Vandermosten et al., 2015).

Summary of Findings in Preschoolers

Taken together, these results suggest that some of the structural and functional brain characteristics observed in children with developmental dyslexia predate the onset of formal reading instruction, indicating that developmental dyslexia may originate from genetically driven altered organization of primarily left-hemispheric cortical areas, but also to the right-hemispheric, temporo-parietal and occipito-temporal regions. These genetic risks interact reciprocally with a multitude of postnatal factors involved in the development of the reading circuitry, causing the neurobiological and behavioral outcomes observed in developmental dyslexia (see Ozernov-Palchik et al., 2016; van Bergen, van der Leij, & de Jong, 2014). More important, these alterations in preschool children can significantly enhance the prediction of later reading outcomes (Bach, Richardson, Brandeis, Martin, & Brem, 2013; Puolakanaho et al., 2007) or even outperform behavioral variables (Maurer et al., 2009).

BRAIN ATYPICALITIES IN INFANTS AT RISK FOR DYSLEXIA

Because most of the earliest MRI studies were conducted in preschoolers, it remains unclear whether the early brain differences in developmental dyslexia are present at birth or develop in conjunction with the key stages of language development. Longitudinal MRI studies beginning in infancy are needed to better understand the emergence of the developmental dyslexia phenotype and the specific brain subsystems involved. Strong evidence for the prenatal origin of some brain atypicalities in developmental dyslexia has been provided by several studies showing atypical neural responses in FHD+ infants within a few days of birth. For instance, alterations in neural responses to speech sounds have been observed using event-related potentials in FHD+ infants (e.g., Lyytinen et al., 2004; Molfese, 2000; van Herten et al., 2008). More important, these alterations differentiated infants subsequently diagnosed with developmental dyslexia from those who developed typical reading skills (e.g., Molfese, 2000). We have shown that the previously observed white matter alterations in the left AF in children at risk for developmental dyslexia (Wang et al., 2016) can be detected in FHD+ at 5–17 months of age (Langer et al., 2015). The DTI data were collected using the natural sleep paradigm developed in our lab (Raschle, Zuk, Ortiz-Mantilla et al., 2012), and infants' language skills were assessed in infancy and at age 4. An AFQ analysis revealed significantly lower FA in the central portion of the left AF for FHD+ compared with FHD– infants. FA in the left AF correlated with expressive language skills and vocabulary at age 4 in a subsequent preliminary analysis (Figuccio, Wang, Liederman, & Gaab, 2016). It is important to note, however, that it remains to be determined which of these infants will subsequently develop developmental dyslexia.

BRAIN ATYPICALITIES IN THE RIGHT HEMISPHERE: MECHANISMS OF COMPENSATION?

Genetic predisposition for dyslexia is not deterministic, and about 50% of children with familial risk for reading problems will develop average or above average reading skills (Snowling, Muter, & Carroll, 2007), but the protective factors and compensatory mechanisms at play are largely unknown (Pammer, 2014). Neuroimaging studies suggest an increased recruitment of right-hemispheric regions in compensated individuals with dyslexia and in response to interventions (e.g., Barquero, Davis, & Cutting, 2014; Chiarello, Lombardino, Kacinik, Otto, & Leonard, 2006). In a DTI study of prereaders from our lab (Wang et al., 2017), FHD+ children who subsequently developed typical reading skills showed a significantly higher rate of FA development in the temporo-parietal segments of the right SLF compared with those FHD+ children who subsequently developed into poor readers, suggesting the emergence of an alternate network for reading in these children. In another study of middle- and high-schoolers, adolescents with dyslexia showed significant correlations between reading improvement within a 2.5 year period and functional activation in right-hemispheric frontal regions as well as FA in the right SLF. Moreover, the activation pattern in the right frontal areas predicted with

92% accuracy which individual children improved their reading scores over a 2½-year time period (Hoeft et al., 2011).

It is not yet clear from these studies whether recruitment of a right-hemispheric network emerges in response to environment and intervention in some children or whether innate characteristics of this network predispose certain children to respond well to environmental enrichment and explicit instruction, which may facilitate the development of a right-hemispheric reading network. Indeed, a study from our lab showed that activation during a phonological task in anterior right-hemispheric structures exhibited a significantly stronger correlation with home literacy variables in FHD+ compared with FHD– children (Powers, Wang, Beach, Sideridis, & Gaab, 2016), which may suggest emerging right-hemispheric networks in children who are genetically at risk in response to certain environmental inputs that act as protective factors. Future studies are needed to determine whether these alterations are present at birth or emerge over time.

INTEGRATING RESULTS FROM HUMAN NEUROIMAGING INTO THE GESCHWIND-GALABURDA HYPOTHESIS: A FIRST ATTEMPT AND NEXT STEPS

The GGH (Geschwind & Galaburda, 1985a, b, c) has been an influential theoretical framework aimed at integrating clinical observations on handedness, learning variants, immune disorders, and male–female differences into a unified neurobiological framework (see Chapter 1). The GGH postulated that developmental dyslexia is a product of atypical neuronal migration and impeded development of the left hemisphere due to genetic risk factors and the prenatal (in utero) environment. The GGH was based on postmortem findings that suggest brain atypicalities in posterior left hemispheric perisylvian regions in individuals with developmental dyslexia (Galaburda, Sherman, & Rosen, 1985) as well as observations of increased prevalence of learning disabilities in left-handed individuals (Geschwind & Behan, 1982). This chapter provided comprehensive evidence that some brain alterations characteristic for developmental dyslexia can be observed as early as infancy and preschool and specifically in children at hereditary risk for developmental dyslexia (which underlines the need to examine the specificity of these characteristics). This evidence supports the pathway between gene alterations and the range of sensorimotor, perceptual, and cognitive characteristics reported for developmental dyslexia proposed by Galaburda and colleagues (Galaburda, LoTurco, Ramus, Fitch, & Rosen, 2006). Specifically, variant function of genes linked to developmental dyslexia affects cortical development by disrupting in utero neuronal migration, axonal growth, and/or synaptic transmission, resulting in atypical cortico-cortical and cortico-thalamic circuits.

The GGH also proposed that in utero changes in testosterone impede left lateralization in developmental dyslexia, resulting in reduced left-hemispheric asymmetry in the perisylvian regions supporting language development. Interestingly, the findings reviewed here suggest fundamental differences in the development of brain regions supporting reading in individuals with dyslexia, which are primarily observed in the left-hemispheric regions but certainly span both hemispheres. One could hypothesize that variant functions in dyslexia susceptibility genes, alternative genes (Hu, Chahrour, & Walsh, 2014), or atypical hormone levels, as suggested by the GGH, trigger the development of these reduced asymmetries, which, in interaction with the environment, could benefit the development of compensatory mechanisms in children at familial risk. More specifically, due to genetic influences, some children with a familial risk may be less lateralized, enabling them to recruit right-hemispheric networks for specific task demands more easily through increased interhemispheric exchange. This is in line with studies showing increased integrity of the posterior corpus callosum (splenium) in individuals with dyslexia (Frye et al., 2008; Hasan et al., 2012; Odegard, Farris, Ring, McColl, & Black, 2009), which may promote increased interhemispheric transfer and the development of alternative (compensatory) networks.

CONCLUSION

Current results suggest early atypical brain development in children with developmental dyslexia, although the sensitivity and specificity of specific brain alterations need to be determined. It has been further demonstrated that early brain atypicalities in developmental dyslexia are bilaterally distributed but exhibit stronger effects in left-hemispheric structures, and there may be a right-hemispheric compensatory network in some children with a familial risk who subsequently develop into good readers. These findings highlight the significance and potential of a developmental focus starting in early infancy to address the etiology and developmental trajectories of developmental dyslexia. A developmental, neurobiological perspective needs to play a crucial role in developing effective early identification strategies and reducing the clinical, psychological, and social consequences of reading failure. Such a perspective could also inform the development of early interventions prior to important brain network pruning and myelination stages. Understanding the complex etiology of developmental dyslexia will be essential to inform and improve the training of teachers, school psychologists, and other clinicians to reliably recognize and optimize the learning contexts for individual learners; this ultimately could lead to personalized education similar to the model of personalized medicine (Butterworth & Kovas, 2013).

REFERENCES

Altarelli, I., Leroy, F., Monzalvo, K., Fluss, J., Billard, C., Dehaene-Lambertz, G., & Ramus, F. (2014). Planum temporale asymmetry in developmental dyslexia: Revisiting an old question. *Human Brain Mapping, 35*, 5717–5735.

Bach, S., Richardson, U., Brandeis, D., Martin, E., & Brem, S. (2013). Print-specific multimodal brain activation in kindergarten improves prediction of reading skills in second grade. *Neuroimage, 82*, 605–615.

Barquero, L. A., Davis, N., & Cutting, L. E. (2014). Neuroimaging of reading intervention: A systematic review and activation likelihood estimate meta-analysis. *PloS One, 9*(1), e83668.

Black, J. M., Tanaka, H., Stanley, L., Nagamine, M., Zakerani, N., Thurston, A.,…Glover, G. H. (2012). Maternal history of reading difficulty is associated with reduced language-related gray matter in beginning readers. *Neuroimage, 59*(3), 3021–3032.

Boets, B., de Beeck, H. P. O., Vandermosten, M., Scott, S. K., Gillebert, C. R., Mantini, D.,… & Ghesquière, P. (2013). Intact but less accessible phonetic representations in adults with dyslexia. *Science, 342*(6163), 1251–1254.

Butterworth, J., & Kovas, Y. (2013). Understanding neurocognitive developmental disorders can improve education for all. *Science, 340*(6130), 300–305. doi:10.1126/science.1231022

Chi, J. G., Dooling, E. C., & Gilles, F. H. (1977). Gyral development of the human brain. *Annals of Neurology, 1*(1), 86–93.

Chiarello, C., Lombardino, L. J., Kacinik, N. A., Otto, R., & Leonard, C. M. (2006). Neuroanatomical and behavioral asymmetry in an adult compensated dyslexic. *Brain and language, 98*(2), 169–181.

Clark, C. A. H., Specht, K., Narr, K. L. , Manis, F. R., Toga, A. W., & Hugdahl, K. (2014). Neuroanatomical precursors of dyslexia identified from pre-reading through age 11. *Brain*, 137(Pt. 12), 3136–3141.

Figuccio, M. J., Wang, Y., Liederman, J., & Gaab, N. White Matter Connectivity of the Corpus Callosum Assessed in Preschoolers Predicts Reading Fluency in School-Age Children [Abstract]. Society for Research in Child Development, Austin, April, 2017.

Frye, R. E., Hasan, K., Xue, L., Strickland, D., Malmberg, B., Liederman, J., & Papnicolaou, A. (2008). Splenium microstructure is related to two dimensons of reading skills. *Neuroreport, 19*(16), 1627–1631.

Gabrieli, J. D. (2009). Dyslexia: A new synergy between education and cognitive neuroscience. *Science, 325*(5938), 280–283.

Galaburda, A. M., LoTurco, J., Ramus, F., Fitch, R. H., & Rosen, G. D. (2006). From genes to behavior in developmental dyslexia. *Nature Neuroscience, 9*(10), 1213–1217.

Galaburda, A. M., Menard, M. T., & Rosen, G. D. (1994). Evidence for aberrant auditory anatomy in developmental dyslexia. *Proceedings of the National Academy of Sciences USA, 91*(17), 8010–8013.

Galaburda, A. M., Sherman, G. F., & Rosen, G. D. (1985). Developmental dyslexia: Four consecutive patients with cortical anomalies. *Reading, 6*, 1–8.

Geschwind, N., & Behan, P. (1982). Left-handedness: Association with immune disease, migraine, and developmental learning disorder. *Proceedings of the National Academy of Sciences USA, 79*(16), 5097–5100.

Geschwind, N., & Galaburda, A. M. (1985a). Cerebral lateralization. Biological mechanisms, associations, and pathology: I. A hypothesis and a program for research. *Archives of Neurology, 42*(5), 428–459.

Geschwind, N., & Galaburda, A. M. (1985b). Cerebral lateralization. Biological mechanisms, associations, and pathology: II. A hypothesis and a program for research. *Archives of Neurology, 42*(6), 521–552.

Geschwind, N., & Galaburda, A. M. (1985c). Cerebral lateralization. Biological mechanisms, associations, and pathology: III. A hypothesis and a program for research. *Archives of Neurology, 42*(7), 634–654.

Goswami, U. (2015). Sensory theories of developmental dyslexia: Three challenges for research. *Nature Reviews Neuroscience, 16*(1), 43–54.

Hasan, K., Molfese, D., Walimuni, I., Stuebing, K., Papanicolaou, A., Narayana, P., & Fletcher, J. (2012). Diffusion tensor quantification and cognitive correlates of the macrostructure and microstructure of the corpus callosum in typically developing and dyslexic children. *NMR in Biomedicine, 25*(11), 1263–1270.

Hoeft, F., McCandliss, B. D., Black, J. M., Gantman, A., Zakerani, N., Hulme, C.,...Gabrieli, J. D. (2011). Neural systems predicting long-term outcome in dyslexia. *Proceedings of the National Academy of Sciences USA, 108*(1), 361–366.

Horwitz, B., Rumsey, J. M., & Donohue, B. C. (1998). Functional connectivity of the angular gyrus in normal reading and dyslexia. *Proceedings of the National Academy of Sciences USA, 95*(15), 8939–8944.

Hu, W. F., Chahrour, M. H., & Walsh, C. A. (2014). The diverse genetic landscape of neurodevelopmental disorders. *Annual Review of Genomics and Human Genetics, 15*, 195–213.

Im, K., Raschle, N. M., Smith, S. A., Grant, P. E., & Gaab, N. (2015). Atypical sulcal pattern in children with developmentaldyslexia and at-risk kindergarteners. *Cerebral Cortex, 26*(3), 1138–1148.

Kostovic, I., & Vasung, L. (2009, August). Insights from in vitro fetal magnetic resonance imaging of cerebral development. In *Seminars in perinatology* (Vol. 33, No. 4, pp. 220–233). Philadelphia, PA: WB Saunders.

Langer, N., Peysakhovich, B., Zuk, J., Drottar, M., Sliva, D. D., Smith, S.,... & Gaab, N. (2017). White matter alterations in infants at risk for developmental dyslexia. *Cerebral Cortex*, 27(2), 1027-1036.

Lyytinen, H., Aro, M., Eklund, K., Erskine, J., Guttorm, T., Laakso, M. L.,...Torppa, M. (2004). The development of children at familial risk for dyslexia: birth to early school age. *Annals of Dyslexia, 54*(2), 184–220.

Maurer, U., Bucher, K., Brem, S., Benz, R., Kranz, F., Schulz, E.,...Brandeis, D. (2009). Neurophysiology in preschool improves behavioral prediction of reading ability throughout primary school. *Biological Psychiatry, 66*, 341–348.

Molfese, D. L. (2000). Predicting dyslexia at 8 years of age using neonatal brain responses. *Brain Language, 72*(3), 238–245. doi:10.1006/brln.2000.2287S0093-934X(00)92287-9 [pii]

Morken, F., Helland, T., Hugdahl, K., & Specht, K. (2017). Reading in dyslexia across literacy development: A longitudinal study of effective connectivity. *NeuroImage*, 144, 92–100.

Myers, C. A., Vandermosten, M., Farris, E. A., Hancock, R., Gimenez, P., Black, J. M.,... Hoeft, F (2014). Structural changes in white matter are uniquely related to children's reading development. *Psychological Science, 25*(10), 1870–1883.

Niogi, S. N., & McCandliss, B. D. (2006). Left lateralized white matter microstructure accounts for individual differences in reading ability and disability. *Neuropsychologia*, 44(11), 2178–2188. doi:S0028-3932(06)00023-6 [pii]10.1016/j.neuropsychologia.2006.01.011

Norton, E. S., Beach, S. D., & Gabrieli, J. D. (2015). Neurobiology of dyslexia. *Current Opinion in Neurobiology, 30*, 73–78.

Odegard, T. N., Farris, E. A., Ring, J., McColl, R., & Black, J. (2009). Brain connectivity in non-reading impaired children and children diagnosed with developmental dyslexia. *Neuropsychologia*, 47(8), 1972–1977.

Ozernov-Palchik, O., Yu, X., Wang, Y., & Gaab, N. (2016). Lessons to be learned: How a comprehensive neurobiological framework of atypical reading development can inform educational practice. *Current Opinion in Behavioral Sciences, 10*, 45–58.

Pammer, K. (2014). Brain mechanisms and reading remediation: More questions than answers. *Scientifica*,.IS 802741, doi:10.1155/2014/802741

Powers, S. J., Wang, Y., Beach, S. D., Sideridis, G. D., & Gaab, N. (2016). Examining the relationship between home literacy environment and neural correlates of phonological processing in beginning readers with and without a familial risk for dyslexia: an fMRI study. *Annals of Dyslexia*, 66(3), 337–360. doi:10.1007/s11881-016-0134-2

Puolakanaho, A., Ahonen, T., Aro, M., Eklund, K., Leppanen, P. H., Poikkeus, A. M.,... Lyytinen, H. (2007). Very early phonological and language skills: Estimating individual risk of reading disability. *Journal of Child Psychology and Psychiatry, 48*(9), 923–931. doi:JCPP1763 [pii]10.1111/j.1469-7610.2007.01763.x

Raschle, N. M., Chang, M., & Gaab, N. (2011). Structural brain alterations associated with dyslexia predate reading onset. *Neuroimage, 57*(3), 742–749.

Raschle, N. M., Lee, M., Buechler, R., Christodoulou, J. A., Chang, M., Vakil, M., Stering, P. L., & Gaab, N. (2009). Making MR imaging child's play—pediatric neuroimaging protocol, guidelines and procedures. *Journal of Visualized Experiments, 29*,1–9.

Raschle, N. M., Stering, P. L., Meissner, S. N., & Gaab, N. (2013). Altered neuronal response during rapid auditory processing and its relation to phonological processing in prereading children at familial risk for dyslexia. *Cerebral Cortex, 24*(9), 2489–2501. doi:10.1093/cercor/bht104.

Raschle, N. M., Zuk, J., & Gaab, N. (2012). Functional characteristics of developmental dyslexia in left-hemispheric posterior brain regions predate reading onset. *Proceedings of the National Academy of Sciences USA, 109*(6), 2156–2161.

Raschle, N., Zuk, J., Ortiz-Mantilla, S., Sliva, D. D., Franceschi, A., Grant, P. E.,…Gaab, N. (2012). Pediatric neuroimaging in early childhood and infancy: Challenges and practical guidelines. *Annals of the New York Academy of Sciences, 1252*(1), 43–50.

Rimrodt, S. L., Peterson, D. J., Denckla, M. B., Kaufmann, W. E., & Cutting, L. E. (2010). White matter microstructural differences linked to left perisylvian language network in children with dyslexia. *Cortex, 46*(6), 739–749.

Saygin, Z. M., Norton, E. S., Osher, D. E., Beach, S. D., Cyr, A. B., Ozernov-Palchik, O.,… Gabrieli, J. D. (2013). Tracking the roots of reading ability: White matter volume and integrity correlate with phonological awareness in prereading and early-reading kindergarten children. *Journal of Neuroscience, 33*(33), 13251–13258.

Snowling, M. J., Muter, V., & Carroll, J. (2007). Children at family risk of dyslexia: A follow-up in early adolescence. *Journal of Child Psychology and Psychiatry, 48*(6), 609–618.

Stanberry, L. I., Richards, T. L., Berninger, V. W., Nandy, R. R., Aylward, E. H., Maravilla, K. R.,... & Cordes, D. (2006). Low-frequency signal changes reflect differences in functional connectivity between good readers and dyslexics during continuous phoneme mapping. *Magnetic Resonance Imaging, 24*(3), 217–229.

van Bergen, E., van der Leij, A., & de Jong, P. F. (2014). The intergenerational multiple deficit model and the case of dyslexia. *Frontiers in Human Neuroscience, 8*, 346. doi:10.3389/fnhum.2014.00346

van der Mark, S., Klaver, P., Bucher, K., Maurer, U., Schulz, E., Brem, S.,... & Brandeis, D. (2011). The left occipitotemporal system in reading: disruption of focal fMRI connectivity to left inferior frontal and inferior parietal language areas in children with dyslexia. *Neuroimage, 54*(3), 2426–2436.

van Herten, M., Pasman, J., van Leeuwen, T. H., Been, P. H., van der Leij, A., Zwarts, F., & Maassen, B. (2008). Differences in AERP responses and atypical hemispheric specialization in 17-month-old children at risk of dyslexia. *Brain Research, 1201*, 100–105. doi:10.1016/j.brainres.2008.01.060S0006-8993(08)00196-0 [pii]

Vandermosten, M., Boets, B., Wouters, J., & Ghesquiere, P. (2012). A qualitative and quantitative review of diffusion tensor imaging studies in reading and dyslexia. *Neuroscience and Biobehavior Review, 36*(6), 1532–1552. doi:S0149-7634(12)00061-9 [pii]10.1016/j.neubiorev.2012.04.002

Vandermosten, M., Vanderauwera, J., Theys, C., De Vos, A., Vanvooren, S., Sunaert, S.,… Ghesquière, P. (2015). A DTI tractography study in pre-readers at risk for dyslexia. *Developmental Cognitive Neuroscience, 14*, 8–15.

Wang, Y., Mauer, M. V., Raney, T., Peysakhovich, B., Becker, B. L., Sliva, D. D., & Gaab, N. (2017). Development of tract-specific white matter pathways during early reading development in at-risk children and typical controls. *Cerebral Cortex, 27*(4), 2469–2485.

Yeatman, J. D., Dougherty, R. F., Myall, N. J., Wandell, B. A., & Feldman, H. M. (2012). Tract profiles of white matter properties: Automating fiber-tract quantification. *PLoS One, 7*(11), e49790.

Ziegler, A., König, I. R., Deimel, W., Plume, E., Nöthen, M. M., Propping, P.,… Remschmidt, H. (2005). Developmental dyslexia–recurrence risk estimates from a German bi-center study using the single proband sib pair design. *Human Heredity, 59*(3), 136–143.

CHAPTER 16

Reading in Children with Developmental Disorders

Heidi M. Feldman

SUMMARY

Many studies of reading disorders focus on children with isolated and specific difficulties in learning to recognize and decipher printed words, which could be labeled *primary reading disorder*. Reading difficulties can also occur in conjunction with other health or neurodevelopmental disorders, which could be labeled *secondary reading disorder*. This chapter focuses on reading skills in children born preterm (PT) as an example of a secondary reading disorder. A meta-analysis compared children born full term (FT) with children born PT and found that children born PT had lower scores on decoding and comprehension, and the gap between the FT and PT groups in comprehension increased with age. They also have poorer language skills, executive function, and parent-rated attentional capacity. After controlling for the child's level of intelligence, we found that prematurity specifically compromised linguistic processing speed, verbal memory, and reading comprehension, all examples of fluid, rather than crystallized, abilities. Several studies have found that individual differences in reading skills and the presence of a primary reading disorder are associated with measures of coherence of long-range white matter fiber pathways in the brain. We found that both decoding and comprehension skills were associated with white matter coherence in both PT and FT groups. The direction of association was positive in the PT group and negative in the FT group, however, which suggest that the underlying neurobiology of reading skills vary in the two groups. The associations in the FT group were exclusively in the left hemisphere; associations in the PT group were in both hemispheres, suggesting a loss of hemispheric specialization in the PT group, in line with the Geshwind-Galaburda Hypothesis (GGH; Geschwind & Galaburda, 1985a, b, c). Future studies should determine if children with secondary reading disorders require different methods of instruction and/or remediation to maximize their reading outcomes.

INTRODUCTION

Many studies on children with poor reading skills conceptualize their reading disorder as an isolated and specific brain-based difficulty in learning to recognize and decipher printed words. Difficulties in reading comprehension and coexisting conditions, such as other learning disabilities and attention deficits, are conceptualized as derivative of or less important than the primary problem of deciphering words in isolation. The former variant of reading difficulty can be labeled dyslexia. I will label it as a primary reading disorder in this chapter.

Children with secondary reading disorders have medical, developmental, or mental health conditions that compromise the ability to read. The size of this population in relation to primary reading disorders is not known. It is becoming increasingly important, however, to understand and treat the reading disorder in these children. Collectively, children with special health care needs constitute 13%–18% of the population of children (Newacheck et al., 1998). Many have survived conditions, such as premature birth or congenital heart disease, which not long ago might have led to early mortality or profound disability. Now, children with special health care needs reach school age, adolescence, and adulthood in relatively good health. Individuals with secondary reading disorders must learn to read well to succeed as adults in this era in which an increasing proportion of jobs require strong literacy skills and postsecondary education.

The study of secondary reading disorders is likely to make broad theoretical contributions to our understanding of reading, including its neurobiological basis. Secondary reading disorders can demonstrate how applicable a framework such as the GGH (Geschwind & Galaburda, 1985a, b, c) is to conditions beyond its original scope. I discuss secondary reading disorders in relation to two components of the GGH: 1) the prevalence of reading disorders will be higher in boys than girls, based on higher levels of prenatal testosterone; and 2) individual variation in reading will be associated with the degree of lateralization of brain structure or function. The study of secondary reading disorders will hopefully also make practical contributions to education and public policy.

CHILDREN BORN PRETERM AS THE PARADIGM FOR SECONDARY READING DISORDERS

Many factors converge to make prematurity an important condition to study from the standpoint of secondary reading disorders. Prematurity is highly prevalent. Approximately 10% of births occur before 37 weeks gestation; 1.5% are born before 33 weeks. Those born at 29–33 weeks are classified as very low gestational age, and those born before 29 weeks are classified as extremely low gestational age. Adverse neurodevelopmental outcomes are common, especially in the very and extremely low gestational age PT group. Approximately 10%–20% of these children develop severe disabilities, including cerebral palsy, visual or hearing impairment, or intellectual disability; 40%–50% of the children without major disabilities develop disorders of language, learning, attention, behavior, or emotional regulation that may occur in isolation or in clusters (Aarnoudse-Moens, Weisglas-Kuperus, van Goudoever, & Oosterlaan, 2009; Bhutta et al., 2002). Although scores on general intelligence testing are below those of FT peers, many children born PT score in the average range and would be expected to participate in general education. Yet, their challenges in the domains of language, literacy, executive function, and attention lead to high rates of special education (Hollomon & Scott, 1998; Roze et al., 2009; Spencer et al., 2008).

The growing body of literature that links adverse outcomes in individuals born PT to the status of white matter pathways in the brain is an additional factor that motivates the study of reading in the PT group. White matter is composed of bundles of axons that link nearby and distant brain regions. It has long been known from postmortem specimens that children born PT who succumbed to their condition had severe cystic lesions in the periventricular white matter. With the advent of magnetic resonance imaging (MRI), white matter injuries were also found in survivors (Volpe, 2009). These injuries result from the susceptibility of the myelin-producing cell line, oligodendrocyte precursors, to hypoxia, ischemia, and inflammation, particularly between 24- and 32-weeks gestational age. After destruction, new precursors populate injured areas. These replacements show arrested development, however, and fail to generate myelin (Back & Miller, 2014). Improvements in neonatal care have led to a decline in the prevalence of large destructive cysts, yet subtle white matter injuries and associated findings persist (Volpe, 2009). Diffusion MRI (dMRI) documents white matter abnormalities in the absence of cystic lesions (Anjari et al., 2007; Bassi et al., 2011) that persist into adolescence (Travis, Adams, Ben-Shachar, & Feldman, 2015) and adulthood (Allin et al., 2011). Other neural injuries have also been identified in children born PT, including smaller brains (Eikenes, Lohaugen, Brubakk, Skranes, & Haberg, 2011; Kesler et al., 2008) and cerebellar injuries (Limperopoulos, Chilingaryan, Guizard, Robertson, & Du Plessis, 2010).

Measures of white matter pathways in the brain assessed using dMRI are associated with reading skills among the general population. Fractional anisotropy (FA) is a dMRI measure of how water diffusion is restricted within a given voxel or set of voxels of imaged brain. White matter has higher FA than gray matter and cerebrospinal fluid. FA within white matter is generally thought to reflect the coherence of white matter pathways, with higher FA indicative of greater coherence. Other factors, such as the density of crossing fibers within a voxel, may complicate interpretation of FA, however (Jones, Knösche, & Turner, 2013). In a now classic study, FA in bilateral temporo-parietal white matter was found to be reduced in adult poor readers compared with controls (Klingberg et al., 2000). Subsequent studies replicated these results in children (Beaulieu et al., 2005; Deutsch et al., 2005; Lebel & Beaulieu, 2009). It is now established that microstructural properties (e.g., FA) of various white matter tracts in the brains of adults and children are associated with individual differences in reading abilities (Ben-Shachar, Dougherty, & Wandell, 2007; Hoeft et al., 2011; Vandermosten, Boets, Wouters, & Ghesquière, 2012; Wandell & Yeatman, 2013).

READING IN CHILDREN BORN PRETERM

Children born PT are at high risk for developing a secondary reading disorder. A quantitative meta-analysis of studies on academic achievement, behavioral functioning, and executive function confirmed children born PT scored 0.48 standard deviations (SD) below FT peers on reading tests and 0.76 SD below on spelling tests (Aarnoudse-Moens et al., 2009). A second meta-analysis (Kovachy, Adams, Tamaresis, & Feldman, 2015) found that the PT group had significantly lower scores than the FT group for both decoding (Cohen's $d = -0.42$) and reading comprehension (Cohen's $d = -0.57$). The group differences significantly increased with age for reading comprehension and trended toward significance for decoding.

A behavior study of children born PT sought to determine whether language and reading skills would be associated with prematurity, independent of the decrements in intelligence quotients also found in PT populations (Lee, Yeatman, Luna, & Feldman, 2011). The sample was a well-characterized cohort of 100 children born PT at ages 9–16 years. Multiple analyses of covariance were used to determine that prematurity made an independent and unique contribution to the variance in scores on a set of six non-overlapping measures of language and reading that included decoding and comprehension, even after consideration of socioeconomic status (SES) and performance or verbal intelligence (see Table 16.1). Subsequent simple regression analyses found that gestational age was a significant predictor of linguistic processing speed, verbal memory, and reading comprehension, after controlling for SES and performance IQ. Prematurity did not contribute to the variance in receptive vocabulary, syntactic comprehension, or decoding. Prematurity seemed to exert an influence on

Table 16.1. Multiple analyses of covariance for measures of language and reading in relation to performance and verbal IQ, socioeconomic status, and prematurity

Models	Covariate and factors	F	partial h²
Covariate	Performance IQ	10.4***	.47
Factors	Socioeconomic status (SES)	3.5**	.23
	Prematurity	3.4**	.22
Covariate	Verbal IQ	17.3***	.59
Factors	SES	2.5*	.17
	Prematurity	2.7*	.19

From Lee, E. S., Yeatman, J. D., Luna, B., & Feldman, H. M. (2011). Specific language and reading skills in school-aged children and adolescents are associated with prematurity after controlling for IQ. *Neuropsychologia, 49*(5), 909; reprinted by permission.

Measures of language and reading included linguistic processing speed, verbal memory, receptive vocabulary, syntactic comprehension, decoding, and reading comprehension.

* $p < .05$; ** $p < .01$; *** $p < .001$

measures that required real-time analysis of verbal and written material as opposed to measures that assessed well-practiced or crystallized functions.

The results of these meta-analyses and the descriptive study highlight that the secondary reading disorder encompasses decoding and comprehension in children born PT. It is also important to note that reading disorders in children born PT are also associated with other learning disabilities, weak attention, and executive function skills (Aarnoudse-Moens et al., 2009). Prematurity is an independent contributor to reading comprehension difficulties.

NEURAL BASIS OF READING IN CHILDREN BORN PRETERM

We set out to describe the pattern of associations between white matter microstructure and reading skills in children born PT. White matter microstructure had been associated with individual differences in reading and related abilities in children and adolescents born PT (Feldman, Lee, Yeatman, & Yeom, 2012; Frye et al., 2010; Mullen et al., 2011). Previous methods did not provide anatomic specificity to locations of association, however. We revisited the issue using dMRI and tractography because tractography provides the greatest anatomical specificity (Travis, Ben-Shachar, Myall, & Feldman, 2016).

We focused our inquiry to white matter tracts that have been associated with reading abilities in previous studies of children and adolescents: 1) superior longitudinal fasciculus (fronto-parietal segment) and arcuate fasciculus, tracts considered to be part of the dorsal stream of the language and reading network involved in mapping sound to articulatory-based representations (Brauer, Anwander, Perani, & Friederici, 2013; Hickok & Poeppel, 2004; Vandermosten et al., 2012); 2) corticospinal tract, a segment that was a predictor of reading skills in kindergarteners (Myers et al., 2014), and another region that showed changes in FA after successful reading remediation (Keller & Just, 2009); 3) the uncinate, a tract that courses between the anterior temporal lobe and the frontal lobe; and 4) inferior longitudinal fasciculus, which with the uncinate is considered to be part of the ventral pathways of the language and reading network involved in mapping sound to meaning (Hickok & Poeppel, 2004) and orthography to phonology and semantics (Vandermosten et al., 2012). We looked for associations with both decoding and reading comprehension.

The mean scores on reading for PT and FT groups were in the normal range and were not statistically different (Travis et al., 2016). This finding indicated that we could not explain any neurobiological group differences solely based on the differences in reading performance. FA in both groups was associated with measures of single-word reading and comprehension in the superior longitudinal fasciculus, the arcuate, cortico-spinal tract, and uncinate (Table 16.2). There were no associations of reading and FA within

Table 16.2. Spearman correlations between mean fractional anisotropy of significant segments along the white matter tract and two reading measures (decoding and reading comprehension) for full-term and preterm groups

	Full term		Preterm	
Tracts	Decoding	Comprehension	Decoding	Comprehension
SLF[a]-Left	−.57*	−.75***	.49*	.53**
SLF[a]-Right			.51**	.46*
Arcuate-Left		−.66**	.59**	
CST[b]-Left	−.66**		.50**	
CST[b]-Right	−.67**		.61***	.52**
Uncinate-Left	−.58**	−.65**		.50**
Uncinate-Right	−.69***			

From Travis, K. E., Ben-Shachar, M., Myall, N. J., & Feldman, H.M. (2016). Variations in the neurobiology of reading in children and adolescents born full term and preterm. *NeuroImage Clinical, 11,* 560; reprinted by permission.

$*p < .05$; $**p < .01$; $***p < .001$

[a]Superior longitudinal faciculus

[b]Cortico-spinal tract

the inferior longitudinal fasciculus. Most important, the correlations were negative in the FT group and positive in the PT group, which suggests important variations in the neurobiology of reading in children born FT and PT (see Table 16.2).

The regions of associations between FA and either decoding or comprehension were overlapping. These associations were found in the superior longitudinal and uncinate fasciculi in the PT and FT group. The substantial overlap in the patterns of associations between decoding and comprehension and FA likely relate to the tasks used in the analyses; the measure of decoding may have triggered comprehension, and the measure of comprehension definitely required decoding.

How do we explain the opposing directions of association in the two groups? FA is a summary measure of microstructure that arises from multiple tissue properties that may vary within a white matter tract, across tracts, and across clinical populations (De Santis, Drakesmith, Bells, Assaf, & Jones, 2014). Several tissue properties are known to influence FA, including the density of crossing fibers, the degree of myelination, and axonal size. These properties can have counteracting effects on FA while still being relevant for rapid transmission of neural information required for reading (Jones et al., 2013). An increase in the number of crossing fibers or variations in axon diameters within the major tracts of the reading network in the FT group could facilitate information exchange among cortical regions and yet lead to decreased FA within the tracts. Children born FT may also achieve efficient communication between distant brain regions via overgrowth of connections, followed by pruning of inappropriate axons

(Yeatman, Dougherty, Ben-Shachar, & Wandell, 2012). Better performance might be associated with fewer but larger axons capable of faster conduction speeds (Arbuthnott, Boyd, & Kalu, 1980; Horowitz et al., 2015) and yet reduced FA (Dougherty et al., 2007).

The positive associations of FA and reading in the PT group suggests that children born PT can achieve similar levels of reading performance via alternative mechanisms. The association of higher FA and reading suggests that the relevant property for this group includes the degree of myelination of axons. If the PT group experienced early white matter damage and axonal loss, then they may have fewer crossing fibers. Better function for them might be associated with myelin accumulation in coherent tracts, leading to higher FA (Fields, 2008).

Summary

This study found that reading skill in healthy PT and FT children is associated with FA of several white matter tracts, which is consistent with previous studies of FT children. The results, however, suggest that different microstructural properties or combinations of properties of these connections are relevant to performance in the two groups. These findings raise the possibility that other clinical groups who experience secondary reading disorders may have other underlying neurobiological patterns associated with reading skills. The impact of different neurobiological factors on reading assessment and treatment has not been well studied.

IMPLICATIONS FOR THE GESCHWIND-GALABURDA HYPOTHESIS

The GGH (Geschwind & Galaburda, 1985a, b, c) addresses both sex differences and asymmetries in brain structure and function.

Sex Differences

Our studies of children born PT did not find sex differences. We suspect that the reason is a small sample size, however. Male sex is associated with high risk at every stage related to PT birth. PT birth is more prevalent among males than females; the proportion of male PT births is higher earlier in gestation and declines with advancing pregnancy (Challis, Newnham, Petraglia, Yeganegi, & Bocking, 2013). Male sex was associated with higher rates of mortality among infants born at the limits of viability, 22–25 weeks (Lee et al., 2010). Boys have higher rates of adverse health and neurodevelopmental outcomes of prematurity; these effects were not related to neonatal factors and seemed to suggest inherent male vulnerability (Peacock, Marston, Marlow, Calvert, & Greenough, 2012). Boys born PT were more likely than girls to have adverse neurodevelopmental outcomes in the toddler years, including moderate to severe cerebral palsy, low scores on mental and physical developmental testing, deafness, or blindness. Male gender was a significant and independent risk factor for neurodevelopmental impairment in multivariate models (Hintz et al., 2006). At the neurobiological level, boys born PT were found to have lower FA in the right uncinate than girls; FA values in that region correlated with both verbal intelligence and receptive vocabulary (Constable et al., 2008). In addition, although thinning of the corpus callosum in children born PT was seen in both sexes, the association of verbal IQ and verbal fluency scores with midsagittal corpus callosum size and midposterior surface area was restricted to the PT boys (Nosarti et al., 2004).

Although the findings of male predominance in this population are consistent with the GGH (Geschwind & Galaburda, 1985a, b, c), the explanations likely go beyond the influence of prenatal testosterone. The fetal genotype influences placental enzyme activities and fetal pituitary adrenal function. Male sex has been associated with proinflammatory forces and placental lesions (Challis et al., 2013). Maternal immune response against the invading interstitial trophoblast may be an initial event leading ultimately to sex differences in neurodevelopmental outcomes (Challis et al., 2013). The possibility that early and complex mechanisms may influence sex differences in outcomes even after birth deserves study, which may refine our views on sex differences.

Asymmetric Brain Structure and Function

Associations of white matter properties and reading in the FT group was associated with FA in regions of the left-hemisphere superior longitudinal fasciculus and arcuate (Travis et al., 2016). This laterality was not found for the other tracts. The dorsal stream of the language and reading network is generally considered more lateralized than the ventral stream. Regions of the right-hemisphere superior longitudinal fasciculus in PT participants were also associated with reading. The findings are consistent with functional imaging studies that suggest loss of the laterality in the PT group (Mullen et al., 2011; Myers et al., 2010) and suggest a loss of laterality in the PT group at risk for secondary reading disorders, which is consistent with the GGH.

IMPLICATIONS FOR EDUCATION

Educators frequently obtain detailed medical histories on children as they enter school, especially children who receive special education services. It is unclear, however, how they use the information in assessment and treatment planning. It is important to determine whether current assessments appropriately identify and characterize children with secondary reading disorders, given that they may have different patterns of abilities than children with primary reading disorders. It is equally critical to establish whether these children require distinctive approaches to reading education or different interventions than children with primary reading disorders. We encourage educators to determine whether remediation of difficulties with decoding should be integrated with interventions for reading comprehension, executive function, and other learning domains. It may be necessary to establish targeted practice guidelines in the future for children with specific conditions and concomitant secondary reading disorders.

IMPLICATIONS FOR PUBLIC POLICY

Although the scientific literature has documented that children born PT are at high risk for secondary reading disorders, these children do not receive any systematic, preventive interventions for early development or reading. The situation is similar for children with other conditions associated with reading disorders, such as neurofibromatosis or congenital heart disease. They often have to fail before they receive help, just like children with primary reading disorders. Interventions for reading disorders are most successful when provided at the youngest ages, so we are often late in offering necessary interventions. If we are going to go all out medically to provide these children with an excellent chance for long-term survival, then we should match those efforts with robust investments in their education.

The No Child Left Behind Act of 2001 (PL 107-110) and its reauthorization as the Every Student Succeeds Act of 2015 (PL 114-95) are designed to raise academic standards and student achievement. Although these laws are admirable in their efforts to raise reading levels in children who have traditionally done poorly, including underrepresented minorities or homeless youth, they fail to acknowledge that all children are unable to learn to read at the same rates, within the same time frame, and using the same instructional methods. The Every Student Succeeds Act of 2015 has shifted responsibilities to the states to write regulations about testing and accountability. States must balance efforts to improve education for all children by recognizing that many factors put reading (and other skills) at risk, including medical, neurological, and mental health conditions that cause secondary reading disorders. Future policies should provide funding and early support services to parents of children at risk for reading disorders and early care and education to build enthusiasm for reading and enhance children's preliteracy skills prior to grade school. Finally, we must invest in schools so that educators are knowledgeable about the scientific foundations of reading and have the educational resources to work with students who need more time, attention, distinctive approaches, or individualized instruction in order to learn to read.

REFERENCES

Aarnoudse-Moens, C. S., Weisglas-Kuperus, N., van Goudoever, J. B., & Oosterlaan, J. (2009). Meta-analysis of neurobehavioral outcomes in very preterm and/or very low birth weight children. *Pediatrics, 124*(2), 717–728.

Allin, M. P., Kontis, D., Walshe, M., Wyatt, J., Barker, G. J., Kanaan, R. A. ,…Nosarti, C. (2011). White matter and cognition in adults who were born preterm. *PLoS One, 6*(10), e24525.

Anjari, M., Srinivasan, L., Allsop, J., Hajnal, J., Rutherford, M., Edwards, A. D., & Counsell, S. (2007). Diffusion tensor imaging with tract-based spatial statistics reveals local white matter abnormalities in preterm infants. *NeuroImage, 35*(3), 1021–1027.

Arbuthnott, E. R., Boyd, I. A., & Kalu, K. U. (1980). Ultrastructural dimensions of myelinated peripheral nerve fibres in the cat and their relation to conduction velocity. *Journal of Physiology, 308*, 125–157.

Back, S. A., & Miller, S. P. (2014). Brain injury in premature neonates: A primary cerebral dysmaturation disorder? *Annals of Neurology, 75*(4), 469–486.

Bassi, L., Chew, A., Merchant, N., Ball, G., Ramenghi, L., Boardman, J.,…Mosca, F. (2011). Diffusion tensor imaging in preterm infants with punctate white matter lesions. *Pediatric Research, 69*(6), 561–566.

Beaulieu, C., Plewes, C., Paulson, L. A., Roy, D., Snook, L., Concha, L., & Phillips, L. (2005). Imaging brain connectivity in children with diverse reading ability. *Neuroimage, 25*(4), 1266–1271.

Ben-Shachar, M., Dougherty, R. F., & Wandell, B. A. (2007). White matter pathways in reading. *Current Opinion in Neurobiology, 17*(2), 258–270.

Bhutta, A. T., Cleves, M. A., Casey, P. H., Cradock, M. M.,…Anand, K. J. S. (2002). Cognitive and behavioral outcomes of school-aged children who were born preterm: a meta-analysis. *Journal of the American Medical Association, 288*(6), 728–737.

Brauer, J., Anwander, A., Perani, D., & Friederici, A. D. (2013). Dorsal and ventral pathways in language development. *Brain Language, 127*(2), 289–295.

Challis, J., Newnham, J., Petraglia, F., Yeganegi, M., & Bocking, A. (2013). Fetal sex and preterm birth. *Placenta, 34*(2), 95–99.

Constable, R. T., Ment, L. R., Vohr, B. R., Kesler, S. R., Fulbright, R. K., Lacadie, C.,…Reiss, A. R. (2008). Prematurely born children demonstrate white matter microstructural differences at 12 years of age, relative to term control subjects: An investigation of group and gender effects. *Pediatrics, 121*(2), 306–316.

De Santis, S., Drakesmith, M., Bells, S., Assaf, Y., & Jones, D. K. (2014). Why diffusion tensor MRI does well only some of the time: Variance and covariance of white matter tissue microstructure attributes in the living human brain. *Neuroimage, 89*, 35–44.

Deutsch, G. K., Dougherty, R. F., Bammer, R., Siok, W. T., Gabrieli, J. D., & Wandell, B. (2005). Children's reading performance is correlated with white matter structure measured by diffusion tensor imaging. *Cortex, 41*(3), 354–363.

Dougherty, R. F., Ben-Shachar, M., Deutsch, G. K., Hernandez, A., Fox, G. R., & Wandell, B. A. (2007). Temporal-callosal pathway diffusivity predicts phonological skills in children. *Proceedings of the National Academy of Sciences of the United States of America, 104*(20), 8556–8561.

Eikenes, L., Lohaugen, G. C., Brubakk, A. M., Skranes, J., & Haberg, A. K. (2011). Young adults born preterm with very low birth weight demonstrate widespread white matter alterations on brain DTI. *Neuroimage, 54*(3), 1774–1785.

Every Student Succeeds Act of 2015, PL 114-95, 20 U.S.C. §§ 6301 *et seq.*

Feldman, H. M., Lee, E. S., Yeatman, J. D., & Yeom, K. W. (2012). Language and reading skills in school-aged children and adolescents born preterm are associated with white matter properties on diffusion tensor imaging. *Neuropsychologia, 50*(14), 3348–3362.

Fields, R. D. (2008). White matter in learning, cognition and psychiatric disorders. *Trends in Neurosciences, 31*(7), 361–370.

Frye, R. E., Hasan, K., Malberg, B., Desouza, L., Swank, P. R., Smith, K. E., & Landry, S. H. (2010). Superior longitudinal fasciculus and cognitive dysfunction in adolescents born preterm and at term. *Developmental Medicine and Child Neurology, 52*(8), 760–766.

Geschwind, N., & Galaburda, A. M. (1985a). Cerebral lateralization. Biological mechanisms, associations, and pathology: I. A hypothesis and a program for research. *Archives of Neurology*, 42(5), 428–459.

Geschwind, N., & Galaburda, A. M. (1985b). Cerebral lateralization. Biological mechanisms, associations, and pathology: II. A hypothesis and a program for research. *Archives of Neurology, 42*(6), 521–552.

Geschwind, N., & Galaburda, A. M. (1985c). Cerebral lateralization. Biological mechanisms, associations, and pathology: III. A hypothesis and a program for research. *Archives of Neurology, 42*(7), 634–654.

Hickok, G., & Poeppel, D. (2004). Dorsal and ventral streams: A framework for understanding aspects of the functional anatomy of language. *Cognition, 92*(1-2), 67–99.

Hintz, S. R., Kendrick, D. E., Vohr, B. R., Kenneth P. W., Higgins, R. D., & NICHD Neonatal Research Network. (2006). Gender differences in neurodevelopmental outcomes among extremely preterm, extremely-low-birthweight infants. *Acta Paediatrica, 95*(10), 1239–1248.

Hoeft, F., McCandliss, B. D., Black, J. M., Gantman, A., Zakerani, N., Hulme, C.,...Gabrieli, J. D. (2011). Neural systems predicting long-term outcome in dyslexia. *Proceedings of the National Academy of Sciences, 108*(1), 361–366.

Hollomon, H. A., & Scott, K. G. (1998). Influence of birth weight on educational outcomes at age 9: The Miami site of the Infant Health and Development Program. *Journal of Developmental and Behavioral Pediatrics, 19*(6), 404–410.

Horowitz, A., Barazany, D., Tavor, I., Bernstein, M., Yovel, G., & Assaf, Y. (2015). In vivo correlation between axon diameter and conduction velocity in the human brain. *Brain Structure and Function, 220*(3), 1777–1788.

Jones, D. K., Knösche, T. R., & Turner, R. (2013). White matter integrity, fiber count, and other fallacies: The do's and don'ts of diffusion MRI. *Neuroimage, 73*, 239–254.

Keller, T. A., & Just, M. A. (2009). Altering cortical connectivity: Remediation-induced changes in the white matter of poor readers. *Neuron, 64*(5), 624–631.

Kesler, S. R., Reiss, A. L., Vohr, B., Watson, C., Schneider, K. C., Katz, K. H.,...Makuch, R. W. (2008). Brain volume reductions within multiple cognitive systems in male preterm children at age twelve. *Journal of Pediatrics, 152*(4), 513–520.

Klingberg, T., Hedehus, M., Temple, E., Salz, T., Gabrieli, J. D., Moseley, M. E., & Poldrack, R. A. (2000). Microstructure of temporo-parietal white matter as a basis for reading ability: evidence from diffusion tensor magnetic resonance imaging. *Neuron, 25*(2), 493–500.

Kovachy, V. N., Adams, J. N., Tamaresis, J. S., & Feldman, H. M. (2015). Reading abilities in school-aged preterm children: A review and meta-analysis. *Developmental Medicine and Child Neurology, 57*(5), 410–419.

Lebel, C., & Beaulieu, C. (2009). Lateralization of the arcuate fasciculus from childhood to adulthood and its relation to cognitive abilities in children. *Human Brain Mapping, 30*(11), 3563–3573.

Lee, E. S., Yeatman, J. D., Luna, B., & Feldman, H. M. (2011). Specific language and reading skills in school-aged children and adolescents are associated with prematurity after controlling for IQ. *Neuropsychologia, 49*(5), 906–913.

Lee, H. C., Green, C., Hintz, S. R., Tyson, J. E., Parikh, N. A., Langer, J., & Gould, J. B. (2010). Prediction of death for extremely premature infants in a population-based cohort. *Pediatrics, 126*(3), e644–650.

Limperopoulos, C., Chilingaryan, G., Guizard, N., Robertson, R. L., & Du Plessis, A. J. (2010). Cerebellar injury in the premature infant is associated with impaired growth of specific cerebral regions. *Pediatric Research, 68*(2), 145–150.

Mullen, K. M., Vohr, B. R., Katz, K. H., Schneider, K. C., Lacadie, C., Hampson, M.,...Ment, L. R. (2011). Preterm birth results in alterations in neural connectivity at age 16 years. *Neuroimage, 54*(4), 2563–2570.

Myers, C. A., Vandermosten, M., Farris, E. A., Hancock, R., Gimenez, P., Black, J. M.,... Hoeft, F. (2014). White matter morphometric changes uniquely predict children's reading acquisition. *Psychological Science, 25*(10), 1870–1883. doi:10.1177/0956797614544511

Myers, E. H., Hampson, M., Vohr, B., Lacadie, C., Frost, S. J., Pugh, K. R.,...Ment, L. R. (2010). Functional connectivity to a right hemisphere language center in prematurely born adolescents. *Neuroimage, 51*(4), 1445–1452.

Newacheck, P. W., Strickland, B., Shonkoff, J. P., Perrin, J. M., McPherson, M., McManus, M.,...Arango, P. (1998). An epidemiologic profile of children with special health care needs. *Pediatrics, 102*(1), 117–123.

No Child Left Behind Act of 2001, PL 107-110, 115 Stat. 1425, 20 U.S.C. §§ 6301 *et seq.*

Nosarti, C., Rushe, T. M., Woodruff, P. W., Stewart, A. L., Rifkin, L., & Murray, R. M. (2004). Corpus callosum size and very preterm birth: relationship to neuropsychological outcome. *Brain, 127*(Pt 9), 2080–2089.

Peacock, J. L., Marston, L., Marlow, N., Calvert, S. A., & Greenough, A. (2012). Neonatal and infant outcome in boys and girls born very prematurely. *Pediatric Research, 71*(3), 305–310.

Roze, E., Van Braeckel, K. N. J. A., van der Veere, C. N., Maathuis, C. G. B., Martijn, A., & Bos, A. F. (2009). Functional outcome at school age of preterm infants with periventricular hemorrhagic infarction. *Pediatrics, 123*(6), 1493–1500.

Spencer, M. D., Moorhead, T. W. J., Gibson, R. J., McIntosh, A. M., Sussmann, J. E. D., Owens, D. G. C.,...Johnstone, E. C. (2008). Low birthweight and preterm birth in young people with special educational needs: A magnetic resonance imaging analysis. *BMC Medicine, 6*(1), 1.

Travis, K. E., Adams, J. N., Ben-Shachar, M., & Feldman, H. M. (2015). Decreased and increased anisotropy along major cerebral white matter tracts in preterm children and adolescents. *PLoS One, 10*(11), e0142860.

Travis, K. E., Ben-Shachar, M., Myall, N. J., & Feldman, H.M. (2016). Variations in the neurobiology of reading in children and adolescents born full term and preterm. *NeuroImage Clinical, 11*, 555–565.

Vandermosten, M., Boets, B., Wouters, J., & Ghesquière, P. (2012). A qualitative and quantitative review of diffusion tensor imaging studies in reading and dyslexia. *Neuroscience and Biobehavioral Reviews, 36*(6), 1532–1552.

Volpe, J. J. (2009). Brain injury in premature infants: A complex amalgam of destructive and developmental disturbances. *Lancet Neurology, 8*(1), 110–124.

Wandell, B. A., & Yeatman, J. D. (2013). Biological development of reading circuits. *Current Opinion in Neurobiology, 23*(2), 261–268.

Yeatman, J. D., Dougherty, R. F., Ben-Shachar, M., & Wandell, B. A. (2012). Development of white matter and reading skills. *Proceedings of the National Academy of Sciences, 109*(44), E3045–E3053.

INTEGRATIVE SUMMARY 4

Reading Skill and the Geschwind-Galaburda Hypothesis

Nicole Landi

SUMMARY

Prior to the availability of neuroimaging or human genotyping, the Geschwind-Galaburda Hypothesis (GGH; Geschwind & Galaburda, 1985a, b, c) identified a path by which gene and environment could interact to produce a brain that would struggle to learn to read. Over the past three decades, advances in both genetic and neuroimaging research have confirmed some of the basic predictions of the GGH and provided new insights into the biological bases of dyslexia. Consistent with the GGH, neuroimaging studies have found atypical development of the left-hemisphere brain regions needed for reading in children with dyslexia. In addition, genetic studies have identified "dyslexia genes" that contribute to the atypical neuronal migration observed by Galaburda and colleagues in the post mortem brains of individuals who had dyslexia. Family studies have recently shown that the size and function of the brain structures needed for reading are more similar among related individuals, and dyslexia-associated left-hemisphere disruptions in the brain are present in young prereading family members who have dyslexia, suggesting a strong genetic component to the neural features that underlie poor reading. On the side of nurture, other research has found that environmental factors súch as low socioeconomic status (SES) and preterm birth also contribute to the atypical brain development associated with poor reading. Ongoing research is utilizing newly available technologies, including whole genome sequencing in combination with state of the art neuroimaging analyses in order to more precisely pinpoint the relative contributions of specific genes and environmental factors to the atypical development of brain regions needed for reading. Although the advanced methods presented in this chapter have led to a greater understanding of the neurobiological bases of dyslexia, they also present an increasing need for specialized training and pose challenges for communicating the significance and nuances of new findings to educators and policy makers.

INTRODUCTION AND SECTION SUMMARY

The GGH (Geschwind & Galaburda, 1985a, b, c) implicated both genetic factors and atypical sensitivity to testosterone during prenatal development in atypical neural development and cerebral lateralization in the dyslexic brain. The GGH identified a path by which gene and environment could interact to produce a complex neurodevelopmental disorder prior to the availability of neuroimaging or human genotyping. The four chapters in this section focus on state of the art methods used since the GGH to identify genetic, neurological, and environmental contributions to dyslexia, or reading disability.

Kornilov and Grigorenko (Chapter 13) begin by noting the problem of missing heritability. Our understanding of the specific genetic mechanisms at work remains largely unknown, despite clear behavioral genetic evidence that reading disability has a strong genetic component. This chapter provides an overview of recent findings from next generation molecular genetic techniques, including genome-wide association studies (GWAS), copy number variant (CNV) association studies, whole-exome and whole-genome sequencing, and genetic approaches that include neuroimaging endophenotypes.

Kornilov and Grigorenko seat these approaches within the context of two models that make different assumptions about the contribution of genes to reading disability. The first proposes a common disorder-common variant (CDCV) approach, which assumes a pattern of polygenic inheritance that includes multiple genetic risk factors that individually contribute small amounts of variance in reading skill but collectively add up to an impaired phenotype. The second is the common disorder-rare variant (CDRV) model, which assumes that each case of reading disability is primarily caused by a single rare variant with a large effect size. Guided by the first model, several recent GWAS with large samples have found significant associations between reading disability phenotypes and previously unidentified candidate genes (e.g., *DAZAP1, CD2L1, ZNF385D, COL4A2*) (Eicher et al., 2013; Luciano et al., 2013). Furthermore, studies that include neuroendophenotypes (using magnetic resonance imaging [MRI] and electroencephalography) have revealed associations between several dyslexia candidate genes and neurobiomarkers associated with reading disability, including gray and white matter structure and patterns of neural response to language stimuli (e.g., Cope et al., 2012; Darki, Peyrard-Janvid, Matsson, Kere, & Klingberg, 2012; Gialluisi, Guadalupe, Francks, & Fisher, in press; Roeske et al., 2011), thus providing specific gene-brain-reading behavior links. Notably, some of the genes identified are involved in neuronal migration, which may contribute to the atypical development of left-hemisphere language regions in reading disability, initially observed by Galaburda and Kemper (1979) and then by Galaburda, Sherman, and Rosen (1985). Furthermore, whole-exome and CNV-association approaches (guided by the CDRV model) provide early suggestive evidence for several rare variants and novel pathways in genes that play a role in cell adhesion and cell division (e.g., *CEP63;* Einarsdottir et al., 2015).

Hoeft and Hancock (Chapter 14) suggest that the study of intergenerational transmission, primarily through comparative neuroimaging within families, can identify neuroendophenotypes, and possible mechanisms of reading disability transmission. Furthermore, given the hypothesized role of factors such as in utero testosterone exposure (per the GGH), the authors highlight the need to consider pre- and postnatal effects as part of intergenerational transmission. They review several methods that have been used to begin to understand the relationship between familial genetic and neurobiological similarity. The first compares siblings who are prereaders with no genetic risk for reading disability to those with familial risk; such studies have revealed several early neurobiomarkers of reading disability, including early disruptions in neural function and structure in key left-hemisphere regions, including temporo-parietal and occipito-temporal (c.f. Vandermosten, Hoeft, & Norton, 2016). A second approach combines a standard twin design with neuroimaging measures and identifies high heritability for cortical thickness, gray matter volume, and white matter integrity in regions that subserve language and reading (c.f. Blokland, de Zubicaray, McMahon, & Wright, 2012). Another approach attempts to tease apart the contributions of gene versus environment for neural structure and function, with some studies finding a significant role for environmental factors such as SES (Noble, Wolmetz, Ochs, Farah, & McCandliss, 2006) and others suggesting a stronger role for genetics (Monzalvo, Fluss, Billard, Dehaene, & Dehaene-Lambertz, 2012). Hoeft and Hancock also discuss the missing heritability problem, which acknowledges the limited contribution of all identified genes to the reading disability phenotype and further motivates the continued exploration of environmental factors, the contribution of de novo mutations, and epigenetic studies of reading disability.

Ozernov-Palchik and Gaab (Chapter 15) focus on the advent of neuroimaging as a new lens for expanding on the GGH, particularly with respect to pediatric imaging. Chapter 15 focuses on two predictions borne out by the GGH: atypical brain development and atypical lateralization in children with reading disability. Like Hoeft and Hancock, these authors note the importance of the postnatal environment in shaping brain development, including the process of learning to read; until recently, the lack of imaging data on infants and young prereaders limited cause-and-effect conclusions that could be drawn from neuroimaging studies of older children and adults with dyslexia. Across studies, school-age children with reading disability present with atypical functional neural activation during reading across a largely left-lateralized network that includes temporo-parietal, occipito-temporal, and inferior frontal regions (c.f. Pugh et al., 2013). In addition, these children show atypical structure in these same regions as well as both functional and structural connectivity among these regions (c.f. Linkersdörfer, Lonnemann, Lindberg, Hasselhorn, & Fiebach, 2012; Maisog, Einbinder, Flowers, Turkeltaub, & Eden, 2008; Rimrodt, Peterson, Denckla, Kaufmann, & Cutting, 2010). Critically, longitudinal and cross-sectional studies of at-risk prereaders have resulted in four key findings: 1) atypical functional activation in many of these same regions

(Raschle, Zuk & Gaab, 2012); 2) correlations of both structure and function with prereading skills (Raschle, Chang, & Gaab, 2011); 3) atypical sulcal patterns (Im, Raschle, Smith, Grant, & Gaab, 2016; and 4) atypical white matter integrity in tracts connecting language and reading-associated areas and correlations between fractional anisotropy (FA) and later reading skill (Wang et al., in press). These findings suggest that atypical neural organization in reading-associated left-hemisphere regions and white matter tracts connecting these regions is likely to be a cause rather than a consequence of reading disability. Gaab, Yu, and Ozernov-Palchik also discuss potential compensatory mechanisms, which have largely involved recruitment of the right hemisphere, including findings that identified early increased white matter connectivity to the right hemisphere in at-risk prereaders who went on to have good reading skills relative to those who went on to be poor readers (Wang et al., in press). Furthermore, new evidence indicates that home literacy variables may contribute to this right-hemisphere compensation in children who are at risk, reiterating the need to consider gene–environment interactive effects (Powers, Wang, Beach, Sideridis, & Gaab, 2016).

Although the prior chapters are primarily concerned with idiopathic reading disability, Feldman (Chapter 16) focuses on children who suffer from reading difficulty and other developmental impairments because of preterm birth. Although not all children born preterm will go on to have developmental disabilities, adverse outcomes are common, especially for those born at a very low gestational age (e.g., Aarnoudse-Moens, Weisglas-Kuperus, van Goudoever, & Oosterlaan, 2009). At school age, preterm children score significantly below their typically developing peers on measures of reading and spelling, and large-scale studies reveal an independent, unique contribution of prematurity to variance in reading ability, even after consideration of other environmental factors such as SES (Lee, Yeatman, Luna, & Feldman, 2011). Postmortem and neuroimaging studies have revealed a potential link between atypical white matter development and negative outcome in preterm infants (c.f. Travis, Adams, Ben-Shachar, & Feldman, 2015), which is also common in children with reading disability as their primary disorder and in prereaders at genetic risk for reading disability (see Chapters 14 and 15). A recent investigation of white matter microstructure (fractional anisotropy [FA]) and reading ability in PT and FT children who had comparable reading proficiency revealed positive associations between reading skill and FA in children born preterm and negative associations in full-term children (Travis, Ben-Shachar, Myall, & Feldman, 2016). Feldman and colleagues interpret this as implicating potentially different neurobiological mechanisms that support reading competency in the two groups of children. Furthermore, correlations between reading and FA in the full-term infants were lateralized to tracts in the left hemisphere, whereas the preterm babies showed bilateral correlations, suggesting differences in lateralization for reading in children born preterm versus full term. This work points to the utility of studying reading in children with known primary insult as a method for elucidating potential causal pathways to reading impairment.

IMPLICATIONS AND FUTURE DIRECTIONS

These four chapters highlight the contributions of neuroimaging and genetic methods to understating the mechanisms that give rise to reading disability. Twenty years of research regarding the neural bases of reading disability identified consistent patterns of left-hemisphere functional and structural anomaly in gray and white matter, as well as compensatory right-hemisphere involvement; these findings are broadly consistent with the atypical neuronal migration and lateralization predictions of the GGH. More recent work indicates that many of these anomalies are observed in prereaders who are at genetic risk for reading disability, suggesting that these neural signatures are, in fact, neurobiomarkers that may be causally related to the development of reading disability (see Chapters 14 and 15). Research discussed in Chapter 13 highlights the utility of modern molecular genetic techniques, including methods for identification of both rare and common variant contributions to reading disability, and the promise of combined gene-brain-behavior approaches for identifying reading disability endophenotypes. Many of the genes identified by early linkage studies and newer methods are implicated in neuronal migration, once again providing continuity with the GGH. Furthermore, the authors of Chapters 14 through 16 discuss the contribution of environmental and other risk factors, including SES, home literacy environment, and preterm birth, which may interact with genetic variables to influence the transmission of reading disability.

The cutting-edge methods and approaches discussed in these four chapters have dramatically changed the landscape of our knowledge of reading disability at every level. Some obstacles remain, however, despite these advances. Rapidly advancing neuroimaging and molecular genetic technologies require greater training and expertise for the scientists who use them and necessitate collaborative team science, which is expensive and often difficult to coordinate. Furthermore, much of the ongoing molecular genetic work is done in isolation, with groups often studying different genes or gene families, in part due to the high-profile impact of new genetic findings, which limits the degree to which comparable work is done simultaneously to boost reproducibility. Although approaches that combine molecular genetics and neuroimaging work (along with careful phenotyping) hold promise for identification of causal chains, particularly for genes with relatively unknown function, this work is expensive and it is often difficult to obtain the large *N*s needed for current-generation genetic analyses.

Several chapters in this volume point to the value of longitudinal imaging research beginning early with prereaders and infants who are at risk; however, this work requires particularly specialized approaches (e.g., getting preschoolers to fall asleep before scanning) and can take years to connect with later outcomes for these children. Moreover, although each chapter in this section alludes to the role of the environment (both pre- and postnatal) in the etiology of reading disability, our understating of the precise way in which nature and nurture interact to produce a reading disability phenotype remains underspecified. Ongoing research must do more to measure environmental variables and consider them in analyses. On the flip side, measuring all potential pre- and postnatal environmental variables is simply not possible—geneticists, imaging scientists, and behavioral researchers need to work together to identify which environmental factors are likely to have the biggest impact and identify robust methods of measurement for these variables. Finally, although substantial efforts have been made to disseminate neuroimaging and genetic findings to educators and policy makers, their value is often difficult to communicate and harder still to translate into education dollars.

REFERENCES

Aarnoudse-Moens, C. S. H., Weisglas-Kuperus, N., van Goudoever, J. B., & Oosterlaan, J. (2009). Meta-analysis of neurobehavioral outcomes in very preterm and/or very low birth weight children. *Pediatrics, 124*(2), 717–728.

Blokland, G. A., de Zubicaray, G. I., McMahon, K. L., & Wright, M. J. (2012). Genetic and environmental influences on neuroimaging phenotypes: A meta-analytical perspective on twin imaging studies. *Twin Research and Human Genetics, 15*(3), 351–371.

Cope, N., Eicher, J. D., Meng, H., Gibson, C. J., Hager, K., Lacadie, C.,…Gruen, J. R. (2012). Variants in the *DYX2* locus are associated with altered brain activation in reading-related brain regions in subjects with reading disability. *NeuroImage, 63*(1), 148–156. http://doi.org/10.1016/j.neuroimage.2012.06.037

Darki, F., Peyrard-Janvid, M., Matsson, H., Kere, J., & Klingberg, T. (2012). Three dyslexia susceptibility genes, DYX1C1, DCDC2, and KIAA0319, affect temporo-parietal white matter structure. *Biological Psychiatry, 72*, 671–676.

Eicher, J. D., Powers, N. R., Miller, L. L., Akshoomoff, N., Amaral, D. G., Bloss, C. S.,… Pediatric Imaging, Neurocognition, and Genetics Study. (2013). Genome-wide association study of shared components of reading disability and language impairment. *Genes, Brain, and Behavior, 12*(8), 792–801.

Einarsdottir, E., Svensson, I., Darki, F., Peyrard-Janvid, M., Lindvall, J., Ameur, A.,… Matsson, H. (2015). Mutation in CEP63 co-segregating with developmental dyslexia in a Swedish family. *Human Genetics, 134*(11-12), 1239–1248.

Galaburda, A. M., & Kemper, T. L. (1979). Cytoarchitectonic abnormalities in developmental dyslexia: A case study. *Annals of Neurology, 6*, 94–100.

Galaburda, A. M., Sherman, G. F., & Rosen, G. D. (1985). Developmental dyslexia: Four consecutive patients with cortical anomalies. *Reading, 6*, 1–8.

Geschwind, N., & Galaburda, A. M. (1985a). Cerebral lateralization. Biological mechanisms, associations, and pathology: I. A hypothesis and a program for research. *Archives of Neurology, 42*(5), 428–459.

Geschwind, N., & Galaburda, A. M. (1985b). Cerebral lateralization. Biological mechanisms, associations, and pathology: II. A hypothesis and a program for research. *Archives of Neurology, 42*(6), 521–552.

Geschwind, N., & Galaburda, A. M. (1985c). Cerebral lateralization. Biological mechanisms, associations, and pathology: III. A hypothesis and a program for research. *Archives of Neurology, 42*(7), 634–654.

Gialluisi, A., Guadalupe, T., Francks, C., & Fisher, S. E. (in press). Neuroimaging genetic analyses of novel candidate genes associated with reading and language. *Brain and Language*. https://doi.org/10.1016/j.bandl.2016.07.002

Im, K., Raschle, N. M., Smith, S. A., Grant, P. E., & Gaab, N. (2016). Atypical sulcal pattern in children with dyslexia and at-risk kindergartners. *Cerebral Cortex, 26*(3), 1138–1148.

Lee, E. S., Yeatman, J. D., Luna, B., & Feldman, H. M. (2011). Specific language and reading skills in school-aged children and adolescents are associated with prematurity after controlling for IQ. *Neuropsychologia, 49*(5), 906–913.

Linkersdörfer, J., Lonnemann, J., Lindberg, S., Hasselhorn, M., & Fiebach, C. J. (2012). Grey matter alterations co-localize with functional abnormalities in developmental dyslexia: An ALE meta-analysis. *PloS One, 7*(8), e43122.

Luciano, M., Evans, D. M., Hansell, N. K., Medland, S. E., Montgomery, G. W., Martin, N. G.,…Bates, T. C. (2013). A genome-wide association study for reading and language abilities in two population cohorts. *Genes, Brain, and Behavior, 12*(6), 645–652.

Maisog, J., Einbinder, E., Flowers, D., Turkeltaub, P., & Eden, G. (2008). A meta-analysis of functional neuroimaging studies of dyslexia. *Annals of the New York Academy of Science, 1145*, 237–259. doi:10.1196/annals.1416.024

Monzalvo, K., Fluss, J., Billard, C., Dehaene, S., & Dehaene-Lambertz, G. (2012). Cortical networks for vision and language in dyslexic and normal children of variable socioeconomic status. *NeuroImage, 61*(1), 258–274.

Noble, K. G., Wolmetz, M. E., Ochs, L. G., Farah, M. J., & McCandliss, B. D. (2006). Brain-behavior relationships in reading acquisition are modulated by socioeconomic factors. *Developmental Science, 9*(6), 642–654.

Powers, S. J., Wang, Y., Beach, S. D., Sideridis, G. D., & Gaab, N. (2016). Examining the relationship between home literacy environment and neural correlates of phonological processing in beginning readers with and without a familial risk for dyslexia: An fMRI study. *Annals of Dyslexia, 66*(3), 227–360.

Pugh, K. R., Frost, S. J., Landi, N., Preston, J. L., Mencl, W. E., & Rueckl, J. G. (2013). Neuroimaging studies of reading disabilities: Toward brain-based learning models. In B. Miller, L. E. Cutting, & P. McCardle (Eds.), *Unraveling reading comprehension: Bavioral, neurobiological, and genetic components* (pp. 167–174). Baltimore, MD: Paul H. Brookes Publishing Co.

Raschle, N., Chang, M., & Gaab, N. (2011). Structural brain alterations associated with dyslexia predate reading onset. *Neuroimage, 57*, 742–749.

Raschle, N. M., Zuk, J., & Gaab, N. (2012). Functional characteristics of developmental dyslexia in left-hemispheric posterior brain regions predate reading onset. *Procedings of the National Academy of Sciences USA 109*(6), 2156–2161. doi:1107721109 [pii]10.1073/pnas.1107721109

Rimrodt, S. L., Peterson, D. J., Denckla, M. B., Kaufmann, W. E., & Cutting, L. E. (2010). White matter microstructural differences linked to left perisylvian language network in children with dyslexia. *Cortex, 46*(6), 739–749. doi:S0010-9452(09)00222-6 [pii]10.1016/j.cortex.2009.07.008

Roeske, D., Ludwig, K. U., Neuhoff, N., Becker, J., Bartling, J., Bruder, J.,…Schulte-Korne, G. (2011). First genome-wide association scan on neurophysiological endophenotypes points to trans-regulation effects on SLC2A3 in dyslexic children. *Molecular Psychiatry, 16*, 97–107.

Travis, K. E., Adams, J. N., Ben-Shachar, M., & Feldman, H. M. (2015). Decreased and increased anisotropy along major cerebral white matter tracts in preterm children and adolescents. *PLoS One, 10*(11), e0142860.

Travis, K. E., Ben-Shachar, M., Myall, N. J., & Feldman, H. M. (2016). Variations in the neurobiology of reading in children and adolescents born full term and preterm. *NeuroImage: Clinical, 11:* 555–565.

Vandermosten, M., Hoeft, F., & Norton, E. S. (2016). Integrating MRI brain imaging studies of pre-reading children with current theories of developmental dyslexia: A review and quantitative meta-analysis. *Current Opinion in Behavioral Sciences, 10*, 155–161.

Vandermosten, M., Vanderauwera, J., Theys, C., De Vos, A., Vanvooren, S., Sunaert, S.,…Ghesquière, P. (2015). A DTI tractography study in pre-readers at risk for dyslexia. *Developmental Cognitive Neuroscience, 14*, 8–15.

Wang, Y., Mauer, M., Raney, T., Peysakhovich, B., Becker, B., Sliva, D., & Gaab, N. (2017). Development of tract-specific white matter pathways during early reading development in at-risk children and typical controls. *Cerebral Cortex 27*(4), 2469-2485.

Epilogue

In this volume, we have brought together authors who are using new methods such as functional neuroimaging and genetics, and other new technologies in their research, to consider what has been accomplished since the publication of the Geschwind-Galaburda Hypothesis (GGH) three decades ago (Geschwind & Galaburda, 1985a-c), and to project future directions for addressing aspects of the GGH that remain open for such investigations. Each of the integrative summaries nicely outline next directions in research and clearly convey that while we have made great progress, a substantial number of questions remain. It is our hope that during the next 30 years researchers will address many of these unanswered research questions, address such issues as sex differences at multiple neurobiological levels, consideration of hormonal influences, and examination of brain networks. Importantly, the GGH set the stage for examination of dyslexia from multiple viewpoints, and the idea that it results from a confluence of biological—and environmental—factors. The next 30 years of research on dyslexia will provide critical insights from multiple viewpoints and methodologies; those findings will help inform ways to better prevent, identify, and treat dyslexia.

Index

Page references followed by *f* or *t* indicate figures or tables, respectively.